Colorado River Basin Water Management
Evaluating and Adjusting to Hydroclimatic Variability

Committee on the Scientific Bases of Colorado River Basin
Water Management

Water Science and Technology Board

Division on Earth and Life Studies

NATIONAL RESEARCH COUNCIL
OF THE NATIONAL ACADEMIES

THE NATIONAL ACADEMIES PRESS
Washington, D.C.
www.nap.edu

THE NATIONAL ACADEMIES PRESS   500 Fifth Street, N.W. Washington, DC 20001

NOTICE: The project that is the subject of this report was approved by the Governing Board of the National Research Council, whose members are drawn from the councils of the National Academy of Sciences, the National Academy of Engineering, and the Institute of Medicine. The members of the panel responsible for the report were chosen for their special competences and with regard for appropriate balance.

Support for this study was provided by the National Research Council's Day Fund, the U.S. Bureau of Reclamation under contract number 05PG303309, the California Department of Water Resources, the Metropolitan Water District of Southern California under contract number 77624, and the Southern Nevada Water Authority. Any opinions, findings, conclusions, or recommendations expressed in this publication are those of the author(s) and do not necessarily reflect the views of the organizations or agencies that provided support for the project.

International Standard Book Number-13: 978-0-309-10524-8
International Standard Book Number-10: 0-309-10524-2

Additional copies of this report are available from the National Academies Press, 500 5th Street, N.W., Lockbox 285, Washington, DC 20055; (800) 624-6242 or (202) 334-3313 (in the Washington metropolitan area); Internet, http://www.nap.edu.

Cover: Tree-ring cross section photograph courtesy of Connie Woodhouse. Glen Canyon Dam and Lake Powell (2004) photograph courtesy of Brad Udall. Skyline photograph of Las Vegas, NV provided by the U.S. Bureau of Reclamation at *http://www.usbr.gov/dataweb/html/ griffith.html*. Formal portrait of John Wesley Powell circa 1890's from the U.S. National Park Service's Grand Canyon National Park Museum Collection.

Copyright 2007 by the National Academy of Sciences. All rights reserved.

Printed in the United States of America.

THE NATIONAL ACADEMIES
Advisers to the Nation on Science, Engineering, and Medicine

The **National Academy of Sciences** is a private, nonprofit, self-perpetuating society of distinguished scholars engaged in scientific and engineering research, dedicated to the furtherance of science and technology and to their use for the general welfare. Upon the authority of the charter granted to it by the Congress in 1863, the Academy has a mandate that requires it to advise the federal government on scientific and technical matters. Dr. Ralph J. Cicerone is president of the National Academy of Sciences.

The **National Academy of Engineering** was established in 1964, under the charter of the National Academy of Sciences, as a parallel organization of outstanding engineers. It is autonomous in its administration and in the selection of its members, sharing with the National Academy of Sciences the responsibility for advising the federal government. The National Academy of Engineering also sponsors engineering programs aimed at meeting national needs, encourages education and research, and recognizes the superior achievements of engineers. Dr. Wm. A. Wulf is president of the National Academy of Engineering.

The **Institute of Medicine** was established in 1970 by the National Academy of Sciences to secure the services of eminent members of appropriate professions in the examination of policy matters pertaining to the health of the public. The Institute acts under the responsibility given to the National Academy of Sciences by its congressional charter to be an adviser to the federal government and, upon its own initiative, to identify issues of medical care, research, and education. Dr. Harvey V. Fineberg is president of the Institute of Medicine.

The **National Research Council** was organized by the National Academy of Sciences in 1916 to associate the broad community of science and technology with the Academy's purposes of furthering knowledge and advising the federal government. Functioning in accordance with general policies determined by the Academy, the Council has become the principal operating agency of both the National Academy of Sciences and the National Academy of Engineering in providing services to the government, the public, and the scientific and engineering communities. The Council is administered jointly by both Academies and the Institute of Medicine. Dr. Ralph J. Cicerone and Dr. Wm. A. Wulf are chair and vice chair, respectively, of the National Research Council.

www.national-academies.org

COMMITTEE ON THE SCIENTIFIC BASES OF COLORADO RIVER BASIN WATER MANAGEMENT[*]

ERNEST T. SMERDON, *Chair*, University of Arizona (Emeritus), Tucson
JULIO L. BETANCOURT, United States Geological Survey, Tucson, Arizona
GORDON W. "JEFF" FASSETT, HDR Engineering, Inc., Cheyenne, Wyoming
LUIS A. GARCIA, Colorado State University, Fort Collins
DONALD C. JACKSON, Lafayette College, Easton, Pennsylvania
DENNIS P. LETTENMAIER, University of Washington, Seattle
ELUID L. MARTINEZ, Water Resources Management Consultants, Santa Fe, New Mexico
STEPHEN C. McCAFFREY, University of the Pacific, Sacramento, California
EUGENE M. RASMUSSON, University of Maryland (Emeritus), College Park
KELLY T. REDMOND, Desert Research Institute, Reno, Nevada
PHILIP M. SMITH, Science Policy and Management, Santa Fe, New Mexico
CONNIE A. WOODHOUSE, University of Arizona, Tucson

NRC Staff

JEFFREY W. JACOBS, Study Director
DOROTHY K. WEIR, Research Associate

---

[*] The activities of this committee were overseen and supported by the National Research Council's Water Science and Technology Board (see Appendix C for listing). Biographical information on committee members and staff is contained in Appendix D.

Preface

The Colorado River has long been uniquely important in the exploration, development, and culture of the western United States. The Colorado is a desert river, stretching from high in the Rockies, through great canyons and arid regions in Utah and Arizona, and finally ending in the Gulf of California in Mexico. For millions of years it has shaped landforms and in the Grand Canyon has exposed geologic formations that are half as old as the Earth itself. The great American scientist John Wesley Powell explored this region widely. He had extensive knowledge of many Native American tribes and his 1869 boating expedition down the Colorado River through the Grand Canyon is legendary. Powell's 1878 publication *Lands of the Arid Region of the United States, with a More Detailed Account of the Lands of Utah* offered many new ideas regarding the roles of the U.S. federal government in developing western water supplies. Although Powell may have foreseen some aspects of western development, one thing he probably did not foresee was the future extent of population growth in the Colorado River region. Nor was Powell likely to have imagined that changes in regional climate might someday affect hydrologic conditions.

Our committee was asked to review the hydrologic and climatic bases of Colorado River water management. In considering this existing body of scientific information, we were struck by the warming across the region in the past century and by the fact that nearly all global climate models forecast increasing temperatures for the Colorado River region. We also noted the exceptionally hot and dry conditions across much of the nation in the summer of 2006, and that the 2006 average annual temperature for the contiguous United States was the warmest on record and nearly identical to the record set in 1998. These conditions are consistent with warming trends in the region.

As we proceeded it became clear that a broad understanding of Colorado River water management issues is not possible unless both water supply and demand issues are adequately considered. Terms such as "population growth" and "water demand" did not appear in our statement of task. As we spoke with water experts from across the region at our meetings, however, they identified important linkages among hydrology and climate and issues such as population growth and water demands, urban water management and conservation, riparian ecology, and water transfers. Clearly, interest in hydroclimatic issues in the region is being driven in large part by increasing water demands and a limited ability to augment water supplies through traditional means. Furthermore, our statement of task called for us to consider the broad topics of systems operations and water management practices. We thus felt it incumbent upon us to comment on topics of water demand, technologies and practices for augmenting water supplies, and programs for coping with drought.

Our report presents population growth data for much of the western United States that is served by Colorado River water. The cities in the region are collectively the fastest growing in the nation. Of further concern is that this growth seems to be occurring with little regard to long-term availability of future water supplies. Ideally, these issues will be openly discussed and squarely addressed before the water supply-and-demand balance across the region becomes more critical. This is important because, for example, the drought of the early 2000s turned out to be even worse than many assumptions regarding a worst-case-scenario drought. This ongoing drought has contained a sequence of exceptionally dry years. Inflows into the basin's storage reservoirs have been well below normal and it may take 15 years of average future hydrologic conditions to refill the basin's largest water storage reservoirs, Lakes Mead and Powell. These hydroclimatic trends are especially troubling in light of rapidly increasing water demands.

I thank our committee members for the hard work and intellect they devoted to producing this consensus report. Each of them brought unique expertise to our deliberations and report preparation and they all devoted many hours of personal time to our study. Their views were fully considered in our study process and I thank them for their contributions, good will, and spirit of collaboration. I also thank the many water scientists, engineers, administrators, and other experts from across the region that spoke with our committee. They provided a comprehensive and fascinating update of key water and science issues across the region and presented important topics and questions for our committee's consideration, all of which were essential to our deliberations and report (Appendix B lists these speakers).

I also thank the National Research Council (NRC) staff members for their dedication and diligent work in our study process. Jeff Jacobs, senior staff officer with the Water Science and Technology Board (WSTB), ensured that our committee stayed on task and that the varying opinions and written contributions from our committee members were blended to create a single, coherent report. Jeff and the committee were ably assisted by WSTB research associate Dorothy Weir, who handled administrative details of the meetings and ably assisted in all phases of report preparation.

We are grateful to the sponsors who provided support for this study. These sponsors included federal, state, and municipal water organizations across the West, which reflects the broad interest in and importance of in these issues. These sponsors were the California Department of Water Resources, the Metropolitan Water District of Southern California, the Southern Nevada Water Authority, and the U.S. Bureau of Reclamation. We also thank the National Academies for providing a substantial portion of funding and for exercising leadership in initiating this study.

This report was reviewed in draft form by individuals chosen for their breadth of perspectives and technical expertise in accordance with the procedures approved by the National Academies' Report Review Committee. The purpose of this independent review was to provide candid and critical comments to assist the institution in ensuring that its published report is scientifically credible and that it meets institutional standards for objectivity, evidence, and responsiveness to the study charge. The reviewer comments and draft manuscript remain confidential to protect the deliberative process. We thank the following reviewers for their helpful suggestions, all of which were considered and many of which were wholly or partly incorporated in the final report: John A. Dracup, University of California; Jerome B. Gilbert, Orinda, California; W.R. Gomes, University of California; Martin P. Hoerling, National Oceanic and

Atmospheric Administration; Malcolm K. Hughes, University of Arizona; Katharine L. Jacobs, University of Arizona; John W. Keys, III, Moab, Utah; Upmanu Lall, Columbia University; John E. Thorson, California Public Utilities Commission; and James L. Wescoat, Jr., University of Illinois.

Although these reviewers provided constructive comments and suggestions, they were not asked to endorse the report's conclusions and recommendations nor did they see the final draft of the report before its release. The review of this report was overseen by Daniel P. Loucks, Cornell University, who was appointed by the NRC's Report Review Committee, and by A. Dan Tarlock, Chicago Kent College of Law, who was appointed by the NRC's Division on Earth and Life Studies. Drs. Loucks and Tarlock were responsible for ensuring that an independent examination of this report was conducted in accordance with NRC institutional procedures and that all review comments received full consideration. Responsibility for this report's final contents rests entirely with the authoring committee and the NRC.

The seven Colorado River basin states and cooperating agencies, particularly the Bureau of Reclamation, face great challenges in addressing the complex issues of Colorado River water supply management. The pressures of meeting the needs of the burgeoning population in the face of future severe droughts and uncertain impacts of global change are indeed great. Political pressures will abound but there are signs of increasing cooperation on a variety of water use issues. We hope this report represents a contribution to the knowledge base of Colorado River hydroclimate and water management and that it helps promote common understanding and cooperation on these matters.

Ernest T. Smerdon
Chair

Contents

SUMMARY	1
1 INTRODUCTION	13
Water Supply Conditions and Hydroclimatic Studies, *17*	
Statement of Task and Scope of Report, *19*	
2 HISTORICAL AND CONTEMPORARY ASPECTS OF COLORADO RIVER DEVELOPMENT	26
Early Exploration and Initial Forays in Colorado River Development: 1860s to 1920, *29*	
Large-Scale Colorado River Water Development: 1920 to 1965, *31*	
Relative Surplus and Shifting Priorities: 1965 to the Mid-1980s, *44*	
Tightening Supplies and Increasing Demands: Mid-1980s to the Present, *48*	
Commentary, *69*	
3 CLIMATE AND HYDROLOGY OF THE COLORADO RIVER BASIN REGION	73
Features and Dynamics of Colorado River Basin Climate, *74*	
Climate Trends and Projections, *80*	
Instrumental Record of Colorado River Streamflow, *92*	
Tree-Ring Science and Reconstructed Streamflow Records, *99*	
Commentary, *108*	
4 PROSPECTS FOR CONSERVING AND EXTENDING WATER SUPPLIES	112
Large-Scale Reservoirs and Inter-Basin Transfers, *113*	
Cloud Seeding, *115*	
Desalination, *118*	
Removing Water-Consuming Invasive Species, *122*	

Agricultural Water Conservation, *123*
Urban Water Conservation, *124*
Offstream Water Banking and Reserves, *127*
Commentary, *131*

5 COLORADO RIVER BASIN DROUGHT PLANNING STRATEGIES AND ORGANIZATIONS   133
Federal-Level Programs, *135*
State-Level Programs, *138*
Municipal-Level Programs, *141*
Other Organizations and Initiatives, *143*
Commentary, *147*

6 EPILOGUE   151

REFERENCES   155

APPENDIXES

A	Letter to Secretary of the Interior Gale A. Norton from the States of Arizona, California, Colorado, Nevada, New Mexico, Utah, and Wyoming Governor's Representatives on Colorado River Operations	175
B	Guest Speakers at Committee Meetings	201
C	Water Science and Technology Board	203
D	Biographical Information for Committee on the Scientific Bases of Colorado River Basin Water Management	205

Summary

Encompassing an area of more than 240,000 square miles, the Colorado River basin covers portions of seven western U.S. states and part of extreme northwestern Mexico. Passing through the heart of what author Wallace Stegner called "the dry core" of the arid western United States, the Colorado's mean annual flow of roughly 15 million acre-feet is not large in comparison to major rivers like the Columbia or the Mississippi. As the largest source of surface water in a large, arid region, however, the Colorado is of great importance to cities, farmers, tribes, anglers, industries, and rafters. In addition to water diversions for municipal, agricultural, and other uses, Colorado River flows generate hydroelectricity, support recreational opportunities and ecological habitats, and sustain cultural and historical values.

Given the Colorado River's importance, variations in its flow record have long been of keen interest to many parties. Direct streamflow measurements date back to the late 1890s when gaging stations were established at a few sites along the river. As the river's flow was measured over the next century, and as a network of stream gaging stations grew, a more complete understanding of Colorado River flows and variability emerged. For example, it is known today that the Colorado River Compact of 1922—the water allocation compact that divides Colorado River flows between the upper and lower Colorado River basin states—was signed during a period of relatively high annual flows. It is also accepted that the long-term mean annual flow of the river is less than the 16.4 million acre-feet assumed when the Compact was signed—a hydrologic fact of no small importance with regard to water rights agreements and subsequent allocations.

Since the 1970s direct measurements of Colorado River flows have been complemented by studies of past hydroclimate conditions that draw from a body of indirect, or proxy, evidence based on tree-ring data. Because patterns of tree-ring growth of trees at lower elevations can reflect moisture availability, tree-ring data can be used to assemble records, or "reconstructions," of past river flows. Using data from coniferous tree species with long life spans in the Colorado River region, flows dating back several centuries have been reconstructed. The first tree-ring-based flow reconstruction for the Colorado River at Lees Ferry, Arizona—the point at which the Colorado River basin is divided legally into its upper and lower basins—was assembled by Charles Stockton and Gordon Jacoby, Jr., in 1976. Additional reconstructions of Colorado River flows that date back to the 15th century, including several undertaken in the past few years, have enhanced scientific understanding of the region's long-term hydrologic and climate patterns.

Tree-ring-based reconstructions became increasingly prominent topics of discussion in western water circles in the early 2000s. Because this period was exceptionally dry across much of the West, the tree-ring-based reconstructions prompted many questions and concerns about the possible extent and severity of future droughts. The water years 2002 and 2004 (as measured from October 1 through the following September 30), for example, were among the 10 driest years of record in the upper basin states of Colorado, New Mexico, Utah, and Wyoming. Significantly, flows into the basin's reservoirs dropped sharply during this period; for example, 2002 water year flows into Lake Powell above Glen Canyon Dam were roughly 25 percent of mean values. These drought conditions stimulated increased interest in tree-ring-based flow reconstructions and long-term Colorado River flows and water availability.

Out of interest in these issues and their implications, in 2005 the National Research Council's Water Science and Technology Board initiated a study to review hydrologic and climatic sciences of the Colorado River region. The Committee on the Scientific Bases of Colorado River Basin Water Management was appointed to assess the extant body of scientific studies regarding both Colorado River hydrology and hydroclimatic trends that might affect river flows. The committee also was asked to consider related topics, including hydrologic models, data, and methods; organizations for evaluating hydroclimatic data; and systems operations and water management practices (the full statement of task to this committee is listed in Chapter 1).

Summary

This committee's statement of task called for a report that produced "an improved hydrologic baseline" for Colorado River water management. In discussing this phrase, the committee noted that it might be interpreted in different ways. An improved hydrologic baseline could, for example, entail a new estimate of long-term mean annual Colorado River flows; establishment of new river gaging stations, computer models, or numerical methods; or a recommendation to reorganize existing (or create new) programs and institutions for evaluating hydrologic and climatic data. After discussing the language in its task statement, the committee concluded that the most appropriate way to help improve a hydrologic baseline for the Colorado River would be to evaluate existing scientific information (including temperature and streamflow records, tree-ring-based reconstructions, and climate model projections) and how it relates to Colorado River water supplies, demands, water management, and drought preparedness.

The following sections of this Summary address the topics of hydroclimatic data and sciences, realities of Colorado River water management, and improving drought preparedness via cooperation, science, and planning. The report's findings and recommendations are presented in bold.

HYDROCLIMATIC DATA AND SCIENCES

Temperature Trends and Model Projections

Temperature records across the Colorado River basin and the western United States document a significant warming over the past century. These temperature records, along with climate model projections that forecast further increases, collectively suggest that temperatures across the region will continue to rise for the foreseeable future. Higher regional temperatures are shifting the timing of peak spring snowmelt to earlier in the year and are contributing to increases in water demands, especially during summer. Higher temperatures will result in higher evapotranspiration rates and contribute to increased evaporative losses from snowpack, surface reservoirs, irrigated land, and vegetated surfaces. Projections of future precipitation are more uncertain than are temperature predictions, leading to uncertainty as to possible changes in future streamflow. Recent studies of the hydrologic implications of warming across the region, based on many global climate models, suggest that on average (across models) runoff and streamflow will decrease. There is, however, uncertainty in these predictions, and some models even suggest increases.

The 20th century saw a trend of increasing mean temperatures across the Colorado River basin that has continued into the early 21st century. There is no evidence that this warming trend will dissipate in the coming decades; many different climate model projections point to a warmer future for the Colorado River region.

Modeling results show less consensus regarding future trends in precipitation. Several hydroclimatic studies project that significant decreases in runoff and streamflow will accompany increasing temperatures. Other studies, however, suggest increasing future flows, highlighting the uncertainty attached to future runoff and streamflow projections. Based on analysis of many recent climate model simulations, the preponderance of scientific evidence suggests that warmer future temperatures will reduce future Colorado River streamflow and water supplies. Reduced streamflow would also contribute to increasing severity, frequency, and duration of future droughts.

Estimating Colorado River Flows: Gaging Stations and Tree-Ring Based Reconstructions

The first gaging stations on the Colorado River were established in the late 19th century. The best-known of the river's many gaging stations is at Lees Ferry, Arizona, established there in 1921. For many years the gaged record of Colorado River flows represented the best science-based knowledge about the river's long-term behavior. Imbedded within this gaged record was an implicit assumption that there was a single, mean value of the river's annual flow, and that interannual variations occurred around this long-term, fixed average. Under this assumption, the basin may have experienced wet and dry periods, but river flows and weather conditions were nonetheless expected to return to an average state, largely defined by climate and hydrology of the early and middle 20th century.

Summary

Questions regarding this long-held paradigm of Colorado River mean discharge arose and have been debated in the latter part of the 20th century. Much of this evolving debate reflected concerns over global climate change that came to prominence beginning in the 1970s. Views of the river's long-term variability continued to evolve with more studies of climate change and hydrology that were conducted beginning in the 1980s. Recent tree-ring-based studies demonstrate that Colorado River flows occasionally shift into decade-long periods in which average flows are lower, or higher, than the 15 million acre-feet/year mean based on the current gaged record. The reconstructions also reinforce the point that the gaged record of Colorado River streamflow covers but a small subset of the range of natural hydroclimatic variability present over several centuries and that future Colorado River hydrology may not mimic the hydrologic behavior reflected within the Lees Ferry gaged record.

For many years, scientific understanding of Colorado River flows was based primarily on gaged streamflow records that covered several decades. Recent studies based on tree-ring data, covering hundreds of years, have transformed the paradigm governing understanding of the river's long-term behavior and mean flows. These studies affirm year-to-year variations in the gaged records. They also demonstrate that the river's mean annual flow—over multidecadal and centennial time scales, as shown in multiple and independent reconstructions of Colorado River flows—is itself subject to fluctuations. Given both natural and human-induced climate changes, fluctuations in Colorado River mean flows over long-range time scales are likely to continue into the future. The paleoclimate record reveals several past periods in which Colorado River flows were considerably lower than flows reflected in the Lees Ferry gaged record, and that were assumed in the 1922 Colorado River Compact allocations.

Tree-Ring-Based Reconstructions, Drought, and Future Water Availability

Tree-ring-based streamflow reconstructions allow the gaged record to be placed in the context of longer-term hydroclimatic variability. Although such reconstructions are only estimates of past river flows, they collectively point to a past in which severe, extended drought was recurrent. They also reveal that 1905-1920 was an exceptionally wet period.

Multicentury, tree-ring-based reconstructions of Colorado River flow indicate that extended drought episodes are a recurrent and integral feature of the basin's climate. Moreover, the range of natural variability present in the streamflow reconstructions reveals greater hydrologic variability than that reflected in the gaged record, particularly with regard to drought. These reconstructions, along with temperature trends and projections for the region, suggest that future droughts will recur and that they may exceed the severity of droughts of historical experience, such as the drought of the late 1990s and early 2000s.

Maintaining the Colorado River Streamflow Gaging Network

The Lees Ferry gage record is an important part of the scientific basis for understanding Colorado River discharge and variability and thus for Colorado River water management. Previous federal-level political and financial support for stream gaging stations has been inconsistent. Over the years, some stations have been discontinued. The loss of stations with long periods (greater than 30 years) of record represents a problem of special concern. The value and importance of reliable and continuous hydrologic records will only grow in the future. It would be imprudent and short-sighted to allow the integrity of the Colorado River gaging station network to be compromised or degraded.

Measured values of streamflow of the Colorado River and its tributaries provide essential information for sound water management decisions. Loss of continuity in this gaged record would greatly diminish the overall value of the existing flow data set, and once such data are lost they cannot be regained. The executive and legislative branches of the U.S. federal government should cooperate to ensure that resources are available for the USGS to maintain the Colorado River basin gaging system and, where possible, expand it.

REALITIES OF COLORADO RIVER WATER MANAGEMENT

In considering its full statement of task and in speaking with Colorado River scientific, engineering, and management experts during the course of this study, this committee identified several trends and realities that affect applications of scientific information in water management. Some of them may prove politically contentious, but they nonetheless merit careful consideration by decision makers at all levels in Colorado River water planning.

Increasing Water Demands, Limited Water Supplies

The late 20th and early 21st centuries witnessed high rates of population growth across the western United States. Population in Arizona, for instance, jumped from about 3.7 million in 1990 to over 5.1 million in 2000—a roughly 40 percent increase (this rate would double Arizona's population in less than 20 years). In Colorado, population grew from slightly fewer than 3.3 million in 1990 to about 4.3 million in 2000—a 30 percent increase. These figures do not necessarily equate directly to increases in water demand; conservation measures, pricing policies, and consumer habits and preferences all influence per capita water uses. In fact, some innovative urban water use and conservation programs have led to reductions in per capita use. Nevertheless, expanding populations have prompted significant increases in urban water demands. Water consumption in Clark County, Nevada (which includes Las Vegas), for example, approximately doubled in the 1985-2000 period. Population growth rates and future projections are on a sharply increasing trajectory in the western United States and they point to sizable and growing water demands for the foreseeable future. In addition, other demands on water supplies, such as those emanating from tribal settlements or from reallocations to support instream flows, will likely grow in the years ahead.

From a water resources perspective, the traditional means of coping with (and effectively encouraging) growth in the western United States was to develop new water supplies by creating large storage reservoirs. After a period of vigorous dam construction in the 1950s and 1960s, prospects for constructing additional large dams in the Colorado River basin have diminished. Today, rather than creating new reservoirs, municipalities are focusing on new, often nonstructural, strategies for augmenting water supplies. A significant trend in this quest has been the sale, lease, and transfer of agricultural water rights to municipalities, particularly in southern California and Colorado. (In Arizona, settlements of tribal water right, with subsequent transfers to municipalities, have also been important.)

Agricultural water rights have been crucial to meeting burgeoning urban water demands in many places. There are other ways for urban areas to obtain additional water supplies, such as through greater use of municipal effluent water (the only growing water supply available in the arid West). Nevertheless, agricultural water appears to constitute the most important, and perhaps final, large reservoir of water available for urban use in the arid U.S. West. In aggregate, the amount of water devoted to agricultural uses is quite large; about 80 percent of western U.S. water supplies are devoted to crop production. Modest shifts of agricultural water to municipal and industrial uses can do much to meet increasing urban water demands. The direct effects associated with the loss of agricultural water, however, such as reduced food production capability, can be significant. In addition, agricultural-urban transfers often entail other "third-party" effects that include costs for rural communities, ecosystems, and other groups indirectly dependent on water supplies affected by the transfers. In recent years many creative water transfer arrangements, involving legally defined water banks and underground water storage programs designed to help mitigate third-party effects, have been developed. The availability of agricultural water is finite, however, and such programs thus are limited in their ability to satisfy increasing, long-term demands. The combination of limited Colorado River water supplies, rapidly increasing populations and water demands, warmer regional temperatures, and the specter of recurrent drought point to a future in which the potential for conflict among existing and prospective new users will prove endemic.

Steadily rising population and urban water demands in the Colorado River region will inevitably result in increasingly costly, controversial, and unavoidable trade-off choices to be made by water managers, politicians, and their constituents. These increasing demands are also impeding the region's ability to cope with droughts and water shortages.

Technologies and Strategies for Augmenting Water Supplies

A wide array of technological and conservation measures can be used to help stretch existing water supplies. These measures include underground storage of water, water reuse, desalination, weather modification, conservation, and changes in water pricing structures and rates. These measures may not necessarily be inexpensive or easy to implement, but many of them show promise and will continue to be pursued and developed as water supplies tighten in future years. Areas experiencing population growth will continue to demand additional water supplies, however, and gains realized through technology, conservation, and other measures will be readily absorbed by increasing population and water demands.

Technological and conservation options for augmenting or extending water supplies—although useful and necessary—in the long run will not constitute a panacea for coping with the reality that water supplies in the Colorado River basin are limited and that demand is inexorably rising.

IMPROVING DROUGHT PREPAREDNESS: COOPERATION, SCIENCE, AND PLANNING

Interstate Cooperation

The drought of the late 1990s and early 2000s prompted the Colorado River states to move toward a new level of interstate cooperation in devising water shortage management criteria. A preliminary proposal presented in a February 2006 letter from the seven basin states to the U.S. Secretary of the Interior (see Appendix A) responded to the Secretary's request that the states develop shortage guidelines and management strategies under low reservoir conditions. This letter represents a noteworthy effort to avoid potential disruptions of operational criteria that govern flow allocations among the basin states.

The interstate cooperation and initiative exhibited by the Colorado River basin states in their February 2006 letter to the Secretary of the Interior is a welcome development that will

prove increasingly valuable—and likely essential—in coping with future droughts and growing water demands.

Scientist-Decision Maker Collaboration

The scientific knowledge base of Colorado River hydrology and climate rivals, and may exceed, comparable knowledge bases for any of the world's river systems. Some of this scientific knowledge has been fundamental to legal and operational decisions, such as the Bureau of Reclamation's Operating Criteria, reservoir operations rule curves, and other aspects of Colorado River basin water resources planning and policy. Some of this scientific information, on the other hand, may not be as well integrated in Colorado River basin water policy as it should be.

Drought conditions in the early 2000s stimulated stronger two-way communication between the scientific community and the water management community. This increased collaboration took the form of workshops, conferences, and other discussions among climate and water experts (especially paleoclimate and tree-ring specialists), hydrologists, civil engineers, and water resources planners and decision makers. Communication between scientists and water managers is important because, for example, it is not always clear what types of scientific information the water management community would find most useful. Scientists can help explain scientific concepts and findings to the water management and user community, while water managers can help scientists frame scientific questions and lines of inquiry that they would find most useful for operational and longer term strategic decisions. These interchanges require sustained, two-way collaboration in order to enhance mutual learning between these groups. It will be important for western water managers to sustain this interest in Colorado River climate, drought, and water planning issues when wetter conditions return, as severe drought conditions will undoubtedly occur again. It will also be important for scientists to sustain their interests in water policy issues related to water supply, demand, and drought management.

A commitment to two-way communication between scientists and water managers is important and necessary in improving overall preparedness and planning for drought and other water shortages. Active communication among people in these commu-

nities should become a permanent fixture within the basin, irrespective of water conditions at any given time. Such dialogue should help scientists frame their investigations toward questions and topics of importance to water managers, and should help water managers keep abreast of recent scientific developments and findings.

Comprehensive, Action-Oriented Study of Pressing Colorado River Water Issues

The Colorado River Compact and much of the Law of the River—the federal and state statutes, interstate compacts, court decisions, and other operating criteria and administrative decisions that define the river's overall governance—were framed during an era in which water for irrigation (and municipal uses in Southern California) was of paramount concern. Today, population growth and increasing water demands have moved urban water issues to the fore of the western water landscape. Increasing urban population and water demands have prompted municipal water managers to think creatively about more efficient water management and ways to increase water supplies and/or limit water use. States and municipalities have sponsored many conservation, landscaping, education, and other related programs. There have been few initiatives, however, to systematically document or synthesize these efforts, which may be hindering progress toward more efficient and better coordinated urban water management across the region. Moreover, knowledge of important topics and issues, such as water demand forecasting and the environmental implications of large-scale agriculture-urban water transfers, lags behind advances in hydrologic and climate sciences.

A more systematic and coordinated approach to urban water conservation and drought preparedness could be promoted through a collaborative investigation across the Colorado River basin. The basin states and municipalities generally establish water practices and policies tailored to their unique circumstances. A comprehensive, accessible report of basin-wide urban water practices, comparing the many lessons learned from diverse experiences across the basin in coping with water shortages and limited supplies, could serve as a more systematic and action-oriented basis for water planning. The collaboration involved in preparing such a report could also promote better communication among federal agencies, the basin states, and municipalities on urban water management strategies and alternatives. It could also encourage a sustained commitment toward a more proactive approach to managing urban water during periods of drought and in the face of growing population.

A comprehensive, action-oriented study of Colorado River region urban water practices and changing patterns of demand should be conducted, as such a study could provide a more systematic basis for water resources planning across the region. At a minimum, the study should address and analyze the following issues:

- historical adjustments to droughts and water shortages,
- demographic projections,
- local and regional water demand forecasting,
- experiences in drought and contingency planning,
- impacts of increasing urban demands on riparian ecology,
- long-term impacts associated with agriculture-urban transfers, and
- contemporary urban water polices and practices (e.g., conservation, landscaping, water use efficiency technologies).

The study could be conducted by the Colorado River basin states, a U.S. federal agency or agencies, a group of universities from across the region, or some combination thereof. The basin states and the U.S. Congress should collaborate on a strategy for commissioning and funding this study. These groups should be prepared to take action based on this study's findings in order to improve the region's preparedness for future inevitable droughts and water shortages.

1

Introduction

The Colorado River basin (Figure 1-1) is renowned for its breathtaking canyons, panoramic natural landscapes, and widespread aridity, as much of the basin lies within the driest region of the continental United States (Figure 1-2). The Colorado River has long intrigued explorers, writers, rafters, and hikers, and the canyons and landscapes shaped by the Colorado and its tributaries have for centuries provided important cultural, social, and spiritual values for many people. The Colorado River's Grand Canyon is one of the nation's great environmental icons and one of the world's favorite tourist destination sites. Place names and sites across the Colorado Plateau—such as Fort Bridger, Disaster Rapids, Crossing of the Waters, Lees Ferry, Monument Valley, Joseph City, Goosenecks of the San Juan, Four Corners, and Mexican Hat—all evoke images of the region's rich history and legend.

With an annual average flow rate of roughly 15 million acre-feet, the Colorado River is not particularly large, especially when compared to U.S. rivers like the Columbia or Mississippi. But the Colorado River is the most important source of water in the vast, arid southwestern United States, providing water for tens of millions of people from San Diego to Denver and a multitude of communities in between. The river thus has been of great interest to hydrologists, water lawyers, municipal water managers, geographers, and civil engineers, and it has been the subject of numerous studies in the fields of physical, natural, and social sciences.

Variations in the Colorado River's flow have long been of interest to water users and managers, and the record of the river's flows based on flow data gathered at Lees Ferry, Arizona, is one of the nation's best-known stream gaging sites. Another noteworthy feature of the Colorado River basin is its large amount of storage capacity relative to the river's flow. The Colorado River system reservoirs have a total of roughly 60 million acre-feet of storage capacity, approximately four times the Colorado's average annual flow. Although the basin's major storage reservoirs have dampened the effects of climate and hydrologic variability, the amount of water in storage remains sensitive to climate fluctuations. Given the strong reliance that steadily increasing populations are placing on the river and its water storage system, variations in Colorado River flows and climate across the basin are as important as they have ever been.

The substantial economic value of Colorado River water has fostered competition—and at times intense animosity—among states and prospective water users. Over the decades, negotiations and legislation involving the Colorado's water resources have added to a considerable body of laws, compacts, treaties, and agreements that partition and allocate its waters; significantly, much of this legal corpus has been designed to accommodate hydrologic and climate variations. Collectively known as the Law of the River, key components of this legal and institutional framework include the 1922 Colorado River Compact, the 1928 Boulder Canyon Project Act, the 1944 Mexico-United States Treaty, the 1948 Upper Colorado River Basin Compact, the Colorado River Storage Project Act of 1956, the 1963 U.S. Supreme Court decision in *Arizona v. California*, the 1968 Colorado River Basin Project Act, the 1973 Minute 242 agreement between Mexico and the United States, the 1992 Grand Canyon Protection Act, and other statutes, court decisions and decrees, contracts, and administrative decisions.

The Law of the River's legal structure is complemented by an elaborate physical infrastructure of dams, reservoirs, levees, canals, aqueducts, tunnels, pumping stations, penstocks, pipes, and ditches. Most of the larger water control structures on the Colorado River and its tributary streams were constructed and today are operated by the U.S. Bureau of Reclamation. The two largest dams across the Colorado River are Hoover Dam, located near Las Vegas, Nevada, and Glen Canyon Dam, located 15 miles south of the Arizona-Utah border. Respectively, these dams impound Lake Mead and Lake Powell, the basin's two primary storage reservoirs. With storage

Introduction

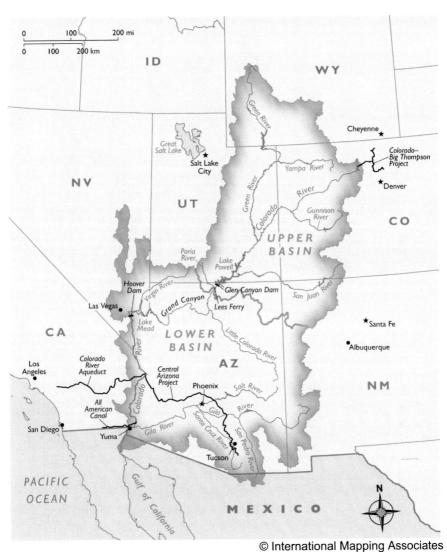

FIGURE 1-1 Colorado River basin.

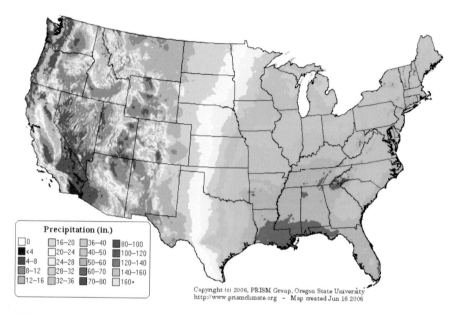

FIGURE 1-2 Average annual precipitation in the United States, 1971-2000.
SOURCE: *http://www.ocs.orst.edu/prism/products/matrix.phtml?vartype= ppt&view=maps.*

capacities of roughly 28 and 27 million acre-feet, including dead storage, each of these reservoirs is capable of storing roughly 2 years of the river's annual mean flow. Storage levels and available capacity of Lake Mead and Lake Powell are particularly important in guiding how the Bureau of Reclamation releases flow through the Colorado River system. Other major facilities and projects within the basin include Flaming Gorge Dam in Wyoming, the Colorado-Big Thompson Project, the Central Utah Project, the Aspinall Unit (which includes Blue Mesa, Crystal, and Morrow Point dams) on the Gunnison River in Colorado, Navajo Dam in New Mexico, the Central Arizona Project's Hayden-Rhodes Aqueduct, the Salt River Project in Arizona, Parker Dam and the Metropolitan Water District of Southern California's Colorado River Aqueduct, Imperial Dam and the All-American Canal serving the Imperial Valley in southern California, and Morelos Dam immediately south of the Mexico-U.S. border. These projects are designed to provide water to users both within and beyond the Colorado River basin, including much of southern California, Colorado's Front Range cities, and the City of Albuquerque.

Introduction

Some of the projects also provide important flood control and re-regulation functions (e.g., storage of water downstream of a dam for additional uses or to further regulate flow).

A prevailing theme in the history of western U.S. water development has been conflict among different users for limited water supplies—especially during drought periods. The legal and physical infrastructure for managing Colorado River water resources was designed to help address or ameliorate conflicts, in part by creating systems to store water during wet periods so that demands during drought can be reliably met. Over much of the 20th century, this system and this water development paradigm generally proved effective at delivering reliable water supplies. Although there have been periods of pronounced drought across the Colorado River basin that caused anxiety (such as during a regional drought in the late 1970s), the large water storage capacity in the basin and the eventual return of wetter conditions heretofore have allowed water delivery obligations to be reliably met.

WATER SUPPLY CONDITIONS AND HYDROCLIMATIC STUDIES

Multiple factors converged across the Colorado River basin during the 1990s and early 2000s that prompted serious concerns among water managers and elected officials regarding long-term water delivery prospects. One of these factors is rapid population growth in urban areas dependent on Colorado River supplies; in particular, Albuquerque, Denver, Las Vegas, Phoenix, Tucson, Los Angeles, and San Diego have all experienced marked increases in population and aggregate water demand since the early 1990s. Beyond fundamental municipal and household uses, these growing urban areas also seek an array of water-related services that includes recreation and instream flow to sustain riparian ecosystems. There have also been settlements of significant tribal water claims, especially in Arizona. This rapidly increasing demand for water poses challenges for water storage systems that are unable to increase supplies by constructing new large-scale storage dams, as was done in the 1950s and 1960s.

Drought conditions that have afflicted much of the Colorado River basin in the late 1990s and early 2000s constitute a second factor driving water supply-and-demand concerns. In the upper Colorado River basin states, the years 2002 and 2004 were among the 10 driest years of hydrologic record. Storage levels in most of the basin's reservoirs dropped markedly during this period—total reservoir system storage in spring 2005 was estimated at roughly 50 percent of average values. Reduced reservoir storage levels, coupled with increasing water demands, have raised the issue of long-term viability of Colorado River water supplies to national-level prominence.

A third factor driving these water supply-and-demand concerns relates to a set of studies from the hydroclimatic sciences community. Hydrologic and climate factors play an important part in Colorado River water supply-and-demand dynamics, and the river's mean annual flow has long been of interest to federal, state, and municipal water managers, tribes, lawyers, users, and scientists. Means for estimating flows are broadly divided into instrumental and pre-instrumental (or "proxy") methods. Colorado River flows are estimated based on data collected by a network of streamflow gages. These instrumental records date back to the late 19th century in some parts of the Colorado River basin. Through the years, gages have been added at other locations, providing a broader picture of the basin's hydrologic flow patterns.

The instrumental record covers roughly one century of Colorado River flow data and is not fully representative of the climatic and hydrologic variability that will occur in the future. Water resources planners thus are interested in other information that might provide a more complete picture of long-term Colorado River flows. A key means of extending the instrumental stream gage record back in time is through proxy techniques that estimate hydrologic data from periods before instrumental techniques were available. Across the western United States and the Colorado River basin, the most prominent proxy technique for investigating past climate involves the study of tree rings, formally known as the science of dendrochronology. The cross sections of coniferous trees exhibit annual growth rings that vary in their thickness, in large part, according to climate variables such as temperature and precipitation. Paleoclimate scientists have long studied tree-ring growth patterns to help deduce past climate conditions. Dendrochronological analyses undertaken over the past 30 years document that severe and extended droughts, significantly impacting Colorado River flows, have occurred many times across the region over the past several centuries. Several of these studies

Introduction

provide estimates of Colorado River streamflows that extend back four to five centuries. An important implication of these studies is that the drought of the late 1990s and early 2000s is hardly unprecedented and that it can readily be understood as part of natural climate variability. Admittedly, in using any proxy methods—including tree-ring reconstructions—one cannot be certain that climate conditions pertaining to proxy indicators represent future (or even current) climate conditions. Nevertheless, these reconstructions identify drought as a recurrent phenomenon, inherent to the region's long-term climate, and that is almost certain to continue to recur in some form in the future. It would be short-sighted to ignore the evidence these proxy methods provide.

In addition to tree-ring-based reconstructions of Colorado River flows, another related body of hydroclimatic studies points out the impacts of increasing temperature on the hydrologic systems of the western United States, the Colorado River basin in particular. Warmer conditions have led to decreased winter snow accumulations and the peak of spring snowmelt runoff is occurring earlier in the year. There is strong evidence that evapotranspiration rates are increasing. Global climate models that project warmer future temperatures—and, in turn, increased rates of evapotranspiration—have important implications for runoff, water storage, and water planning decisions.

STATEMENT OF TASK AND SCOPE OF REPORT

These water supply and demand issues prompted the National Research Council (NRC) in 2005 to undertake this study of the Scientific Bases of Colorado River Basin Water Management (see Box 1-1). The study was carried out by a committee charged to review existing climate and hydrologic studies, methods of data collection and analysis, organizations for managing scientific information, and implications of scientific knowledge for long-term water deliveries and other provisions of the Law of the River. Thus, in addition to reviewing hydrologic and climatic sciences as they pertain to the Colorado River, this study examines linkages between hydroclimatic sciences, and water system operations and management.

> **BOX 1-1**
>
> **Committee on the Scientific Bases of
> Colorado River Basin Water Management**
>
> **Statement of Task**
>
> This activity will assess the extant body of scientific data and studies regarding Colorado River hydrology, including paleohydrological and dendrochronological studies. In addition to paleoscience, the study will also consider other hydroclimatic trends that might influence future hydrologic variability across the river basin. The study's overarching objective will be to help produce an improved hydrologic baseline to be used in support of water project operations and water resources management decisions (e.g., storage operations and diversions) across the Colorado River basin, and other regions of the western United States, especially during periods of extended drought.
>
> These issues will be explored in multiple workshops, to be convened in the region, that will explore the scientific basis of Colorado River hydrology and the strengthening of institutional and related means for gathering and evaluating scientific information. The committee is also likely to convene a meeting, which would be closed to the public, to focus on finalizing its draft report. There are two components to the statement of task for this activity: science and technology, and science institutions and water management practices.

Given the many dimensions of Colorado River scientific and water management issues and the broad statement of task assigned to this committee, a few comments about the scope of this report are in order. The statement of task defines a study that conceivably could cover a vast intellectual area, as it not only calls for a review of existing hydroclimatic studies and data but also mentions modeling techniques and capabilities, decision support systems, institutional arrangements for information gathering and processing, systems operations, water management practices, and water delivery obligations and other relevant dimensions of the Law of the River. Adequate treatment of any of these topics would pose significant challenges to any study group.

Introduction

> 1. Science and Technology
> (a) "Extending" the Colorado River hydrologic record. Historical data, hydrologic and climatic reconstruction methods, and models of Colorado River streamflow reconstruction and hydrology will be reviewed, with a goal of deriving recommendations to strengthen the basis of a synthetic Colorado River streamflow history that more fully reflects long-term conditions than the 100 years of recorded data.
> (b) Hydrologic data, models, and methods. To help advance hydrologic understanding and modeling capabilities in the Colorado River basin, the study will provide advice regarding future research needs and priorities in the realm of hydrologic data availability (i.e., adequacy of the existing stream gaging network) and analysis, hydrologic modeling decision support systems, and related methods.
>
> 2. Science Institutions and Water Management Practices
> (a) Institutional arrangements for establishing a process for gathering and evaluating hydro-climatic variability and water availability information across the Colorado River basin will be explored. The goal is to promote the development and use of a common scientific knowledge base by the basin's numerous water management authorities and entities.
> (b) Systems operations and water management practices. The hydrologic data being evaluated will be examined in the light of its implications for both near-term (e.g., 10 years) water project operations in the basin, and for longer-term water delivery obligations and other relevant dimensions of the Colorado River basin's Law of the River.

Any study of the future of Colorado River water supplies must necessarily be bounded. The need to do so starts with the geographic limits of the Colorado River basin, along with areas beyond the basin served by Colorado River water supplies. The committee recognized and appreciated the numerous water-related issues within and beyond the basin that conceivably could have been addressed in this study; for example, restoration of the Colorado River Delta region, or the potential of offline dams and reservoirs in California to improve water management. In its meetings and discussions, the committee noted a wide variety of important water-related topics—many of which are mentioned in this report—that could have merited more detailed investigation:

- the economics of developing and using alternative water sources; for example, desalination for urban water supplies and wastewater reuse for land applications;
- third-party impacts of water transfers;
- improved agriculture irrigation efficiencies and shifting to higher value crops;
- demand forecasting and management options and demand/price sensitivities;
- more detailed analysis of impacts of temperature increases throughout the basin;
- new technologies for increasing water use efficiencies and their potential impacts;
- impact of droughts on water management decision making in the basin;
- institutional changes needed for more effective integrated water management, adaptive management, real-time monitoring and management, sustainable water development, and so on;
- ways of improving groundwater basin management; and
- offstream water banking.

All these topics are important and conceivably merit a separate volume on their own. Within the scope of its statement of task and its available resources, however, the committee chose to focus on reviewing existing scientific knowledge of hydroclimatic variability and on discussing the implications of hydroclimatic variability in the context of key water management challenges in the Colorado River basin.

With regard to the science and technology portion of its task statement, the committee discussed how to interpret and comment on the phrase an "improved hydrologic baseline." The committee decided that, given its interdisciplinary membership and time and resource limitations, the most appropriate way to approach this phrase (and its statement of task) would be to broadly assess key Colorado River scientific issues as they relate to water supply, demand, management, and drought preparedness. Given the interests in the impli-

Introduction

cations of tree-ring-based reconstructions of Colorado River flows, hydrologic and climate issues in general, and the scientific expertise of the committee membership, it was decided to place some emphasis on these scientific issues (Chapter 3 of this report). The committee also felt it was important to discuss options that might be used to extend water supplies (Chapter 4), and organizations and strategies for improving drought preparedness (Chapter 5).

With regard to the topics of hydrologic models, decision support systems, and institutional arrangements for evaluating hydroclimatic information, the study committee visited with several experts from federal agencies with responsibilities in these areas—namely the Bureau of Reclamation, the National Oceanic and Atmospheric Administration (NOAA), NOAA's National Weather Service, and the U.S. Geological Survey (USGS; the committee included members with experience working within NOAA and the USGS). We spoke with representatives and experts from the seven basin states, from the U.S.-Mexico International Boundary and Water Commission, and with private-sector consultants (Appendix B lists all speakers who visited with this committee). These experts provided an abundance of valuable information, sharing both research findings and personal points of view on Colorado River water and climate issues. In speaking with these experts and in considering current institutional resources and arrangements devoted to evaluating hydroclimate information, it was not clear that any specific new institution, new interagency program, or computer model would represent a notable breakthrough in managing the Colorado River. It also was not clear that significant shortcomings exist within the current arrangements for evaluating Colorado River region hydroclimatic information. A new institution or new arrangement of institutional responsibilities could conceivably lead to greater efficiencies; the institutional analysis required to arrive at such a conclusion, however, was beyond this committee's resources and not consistent with its inclination to focus on science-based topics. Regarding the clause in its charge that refers to the Law of the River, this committee interpreted this as providing latitude to comment on a broad spectrum of water availability and delivery issues as it saw fit. The committee, however, did not evaluate the Law of the River to determine if, for example, laws or treaties might be added or adjusted to better cope with changes in river discharge and new hydroclimatic information. Finally, regarding development and use of a "common scientific knowledge base," this report's final recommendation offers advice on how to strengthen this knowledge base and promote a more systematic basis for water planning across the region.

This report represents the most recent chapter in a series of NRC reviews of Colorado River water management issues that date back to the 1960s. A 1968 report from the NRC Committee on Water, entitled *Water and Choice in the Colorado River Basin: An Example of Alternatives in Water Management*, considered water management approaches and alternatives in the face of growing population and increasing affluence (NRC, 1968). Colorado River water management topics examined in that report, even though it was published nearly 40 years before this report was issued, exhibit some parallels with today's pressing issues. That 1968 report, for example, noted that "in the United States a growing population, increasing affluence, and expanding industry have put new demands on water resources" (NRC, 1968). It asserted the importance of having better information for planning for extended periods of low flow. It also considered prospects for weather modification, desalination technologies, and the use of recycled waste water. Since that 1968 study, the NRC has issued several reports reviewing the operations and downstream environmental impacts at Glen Canyon Dam (NRC, 1987; 1991a; 1996; 1999). These latter reports focus on plans of the U.S Bureau of Reclamation and its efforts to operate the dam to reduce impacts on downstream resources. These NRC reports have been important, for example, in encouraging a more adaptive management regime for Glen Canyon Dam and the Colorado River (e.g., NRC, 1996). The NRC also convened workshops and issued proceedings on the issues of climate change in the western United States (NRC, 1991b) and Colorado River ecology and dam management (NRC, 1991a).

This report should be of interest to a broad audience within the Colorado River basin and to water managers, scientists, scholars, and analysts in other parts of the United States. This audience includes congressional representatives, state legislatures and public officials, and state engineers and water resources and economic planners in the seven Colorado River basin states; federal- and state-level scientists, including hydrologists and climatologists; academic experts with interests in water-related fields; Colorado River basin tribal groups and their representatives; land and natural resources managers; municipalities, farmers, and ranchers that receive deliveries of Colorado River water; environmental groups; hydroelectric power generation organizations and their respective power distributors; and recreational groups such as anglers, kayakers, and rafters that depend on Colorado River water.

This report is organized into six chapters. Following this introductory chapter, Chapter 2 reviews historical events as well as laws, agreements, compacts, and judicial rulings affecting the use of Colorado River water. Chapter 3 discusses important climatic and hydrologic features, data, and methods that underlie the scientific bases of Colorado River water management decisions. Chapter 4 reviews the prospects for extending water supplies via several possible technical and other means. Chapter 5 reviews prominent organizations and programs in the Colorado River region focused on drought detection, response, and mitigation. Chapter 6 offers final reflections on the contents of this report and on adjusting to aridity in the western United States.

2

Historical and Contemporary Aspects of Colorado River Development

Draining an area of over 240,000 square miles, the Colorado River and its main tributary streams originate high within the mountains of western Wyoming, central Colorado, and northeastern Utah. With snowpack accumulating as high as 14,000 feet above sea level, the mainstem of the upper Colorado River receives large amounts of snowmelt from several major tributaries: the Green River flowing south out of Wyoming; the Duchesne River in northern Utah; the Dolores, Gunnison, White, and Yampa rivers in Colorado; and the San Juan River flowing northwest through New Mexico. When it reaches the Canyonlands region of southern Utah (site of Lake Powell), the Colorado's streambed lies hundreds of feet below the surrounding mesas and plateaus. After crossing the Utah-Arizona border and passing Lees Ferry, the river flows westward through Grand Canyon National Park. A further 160 miles downstream—after receiving flows from the Virgin River that drains southwestern Utah and parts of southern Nevada—the Colorado reaches Boulder and Black canyons (which rim much of Lake Mead) and forms the Arizona-Nevada border. Turning southward, the center of the streambed forms the 200-mile-long border between California and Arizona. Near the southern edge of this border, the Gila River (which, along with its tributary the Salt River, drains most of central and southern Arizona) enters the lower Colorado from the east. Just below its confluence with the Gila, the Colorado River enters the state of Sonora, Mexico. There, most of the Colorado's remaining flow is consumed by irrigated agriculture, leaving little water to reach the Gulf of California through the Colorado's historically expansive delta (USBR, 1947; Waters, 1946).

With an annual mean discharge of about 15 million acre-feet, the Colorado River is not a giant among the world's rivers. The Colorado River traverses one of North America's driest regions, however, thus offering opportunities for economic development and growth unmatched by any other water source in this arid region. For the past 100 years these possibilities have spurred myriad political contests among irrigators, businesses, civic boosters, politicians, tribes, ranchers, government officials, engineers, and, more recently, environmental groups and recreational users, all seeking a voice in Colorado River allocation decisions. A root cause of these conflicts is the hydrologic reality that, although roughly 90 percent of the river's flow originates in the upper basin states of Colorado, New Mexico, Utah, and Wyoming, much of the demand for the river's water emanates from the lower basin states of Arizona, California, and Nevada (Hundley, 1966, 1975; Martin, 1989; Moeller, 1971; Pearson, 2002; USBR, 1947).

The story of the development, management, and use of the Colorado River was initially one where concerns over unreliable water supplies were resolved by technological advances, accompanied by legal and administrative arrangements. More recently, this story reflects the concerns of the federal government, the basin states, tribes, municipalities, and other major water users adapting to conditions not fully anticipated when the legal regime and the major dams were put in place.

In the early 20th century, the sparsely populated and largely rural upper basin states watched Southern California's rapid agricultural and urban growth with trepidation. Trepidation turned to fear in 1922 when the Supreme Court held that the western doctrine of prior appropriation could govern apportionment of interstate streams in the arid West. Soon thereafter, the upper basin states succeeded in negotiating the first interstate compact to allocate flows in an interstate stream. The 1922 Colorado River Compact divided the river between the upper and lower basins and reserved unused water for future development in the four upper basin states. Beginning in 1922, California led the fight for the construction of a multipurpose dam on the lower Colorado (decades later they found that the price for having Hoover Dam constructed was a federal apportionment of the river among the three lower basin states). During World War II, political considerations led to a treaty that guaranteed Mexico a supply and in 1948 the upper basin states agreed to an allocation formula among themselves. Once a legal regime (often referred to as "The Law of the River") was in place governing Colorado River water allocations, Congress supported construction of dams on the mainstem and tributaries to support the states' compact rights and delivery obligations. This regime has permitted the basin and major, nearby urban centers—such as Albuquerque, Denver, Los Angeles, Salt Lake City, and San Diego—to grow, but in recent decades it has become stressed by several factors. These include the accommodation of Indian claims, rapid population growth (especially in Arizona and in southern Nevada), the need to control downstream salinity caused by irrigation runoff, disturbances to the Grand Canyon ecosystem caused by the operation of Glen Canyon Dam, and interests in restoring a remnant of the Colorado River Delta in Mexico. These stresses are occurring in the face of the long-standing recognition that the flow estimates on which allocations were negotiated in the 1920 were based upon data drawn from a relatively short and very wet period, and thus turned out to be overly optimistic. Moreover, changes in regional climate conditions may further reduce net available water supplies.

Variations in climate and river flows have been an integral part of this history of Colorado River development. The gathering and analysis of hydroclimatic data assume economic significance because, across the basin, hydrology and climate are linked to larger legal constructs and water development projects. Moreover, the implications of climate and hydrologic studies are related to demographic, water use, and other social and management trends. In reviewing key Colorado River legal agreements and treaties, the history of dam and water storage projects, and demographic and other trends affecting the basin, this chapter does not seek to present an exhaustive discourse; rather, it provides a demographic and legal context for appreciating the significance of subsequent discussions involving climate studies, hydrologic records, water use technologies and practices, and adjustments to drought. This chapter explores this history of the past 150 years or so of Colorado River water development in greater detail. It divides this period into four broad phases: (1) the 1860s through 1920, (2) 1920 to 1965, (3) 1965 to the mid-1980s, and (4) the mid-1980s to the present.

EARLY EXPLORATION AND INITIAL FORAYS IN COLORADO RIVER DEVELOPMENT: 1860s TO 1920

This report focuses on Colorado River development from roughly the middle of the 19th century until the present. Prior to this period there was a rich and extensive history of exploration, irrigated agriculture, and related means for coping with drought and aridity. Spanish explorers led by Coronado in 1540, as well as other expeditions and individuals, referred to the river as the "Colorado" in reference to the reddish silt that—before construction of storage dams—was suspended in the stream's lower reaches. Irrigation in the southwestern United States dates back several centuries to the Hohokam of southern Arizona, who cultivated fields in what is now the greater Phoenix metropolitan region. Spanish settlers, especially in present-day northern New Mexico, later established *acequia* (ditch) systems for irrigation in the 1700s; many of these are still in use. For purposes of this report, discussions of contemporary water management and scientific issues related to the Colorado River basin date back to the 19th-century origins of Anglo-American irrigated agriculture, and to the growth of urban water demand initiated by Los Angeles in the early 20th century. Well before this period, there was an extensive prehistory of water use in the basin, which is chronicled in a substantial body of archaeological and ethnohistorical research (see Brooks, 1974; Dart, 1989; Fish and Fish, 1994; Meyer, 1984). Although a review of long-term social processes dating back several centuries is beyond this report's scope, this body of knowledge could be a valuable resource in helping water managers better cope with hydroclimatic variability. It could be used, for example, in scenario construction, water conservation practices (e.g., reviewing past water harvesting techniques), and forecasting by analogy (see Glantz, 1988).

From the mid-19th century through 1920, the Colorado River basin saw both Anglo-American exploration and the inception of large-scale irrigated agriculture. In the 1860s the upper Colorado River basin constituted one of the last great unexplored regions of North America. Explorer and scientist John Wesley Powell led two important expeditions through this region, the first in 1869 down the Colorado River through Grand Canyon, and the second 2 years later. Boosted by a popular self-penned account of Powell's expeditions, by the late 19th century the Colorado watershed—or at least that encompassing the Grand Canyon and Utah's Canyonlands—had attained almost mythic status in the minds of many Americans (Powell, 1895; Stegner, 1954; Worster, 2000).

The period from the mid-1860s to 1920 witnessed the diffusion of many new irrigation systems throughout the Colorado River basin. In the 1860s Mormon farmers were cultivating fields with water from the Virgin River and, in central Arizona, major irrigation diversions from the Salt and Gila rivers were under way by the end of the decade. In the 1870s farmers began diverting lower Colorado River water for irrigation near Blythe, California, and in the 1880s farmers near Grand Junction, Colorado, were using upper Colorado River flows to nourish crops. These early diversions were relatively minor compared with later development but they established an important precedent that demonstrated future economic and agricultural possibilities (Hundley, 1975; Kleinsorge, 1941; Raley, 2001; Zarbin, 1984, 1997).

Plans for the first major diversion of the Colorado River began in the late 1890s. During this period the California Development Company launched an ambitious plan to divert Colorado River water from near the Mexico-U.S. border and convey it more than 50 miles west to a remote part of Southern California known to 19th-century geographers as the "Colorado Desert." Company boosters changed the region's name to the more inviting "Imperial Valley" and set out to create an agricultural empire encompassing several hundred thousand acres. Imperial Valley irrigation offered enormous possibilities because (1) much of the valley was below sea level, (2) as much as 3 million acre-feet of water could be taken annually from the Colorado River to support irrigation, and (3) in ancient times a channel of the Colorado River—the "Alamo River"—had carried water into the valley. This latter factor proved particularly important because the Alamo Canal of the California Development Company largely followed the ancient channel formed by the Alamo River—thus necessitating little new (and expensive) excavation. A downside to the project (at least in the eyes of many investors and farmers) was that the company's Alamo canal extended through Mexican territory for 50 miles before crossing the international border back into the United States (De Stanley, 1966; Hundley, 1992; Starr, 1990).

By 1900 the California Development Company was delivering water to the Imperial Valley and thousands of settlers were flocking to the region. In 1904 the upper end of the Alamo Canal was reconfigured to counter problems with silt accumulation; unfortunately for the company, in 1905 this canal's newly built wooden headgate was overwhelmed by heavy floods. For the next 2 years the entire flow of the Colorado River descended into the Imperial Valley, drowning crop land and creating a large new waterbody—the Salton Sea—which still exists today. In 1907 the canal heading was finally closed off through laborious efforts of the Southern Pacific Railroad and the flooding stopped, but not before the California Development Company lay in financial ruin. After the company's remaining assets passed to the newly formed Imperial Irrigation District in 1911, local farmers began soliciting federal government support for (1) a flood control dam across the Colorado River to prevent a recurrence of the 1905-1907 disaster, and (2) construction of an "all-American" canal that could deliver Colorado River water to the Imperial Valley without passing through Mexico. Intense lobbying for what eventually became the Boulder Canyon Project Act was under way by 1920 (De Stanley, 1966; Hundley, 1975; 1992; Starr, 1990).

LARGE-SCALE COLORADO RIVER WATER DEVELOPMENT: 1920 TO 1965

Planning for Boulder (later Hoover) Dam in the early 1920s marked the beginning of the second period of Colorado River development; the completion of Glen Canyon Dam in 1964 signaled its end. These two dams comprise the centerpiece of the U.S. federal Colorado River water storage infrastructure. The years 1920-1965 saw a dramatic rise in the influence and prestige of the U.S. Bureau of Reclamation, and most of the basin's large dams were either planned or constructed during this period. Termed the "Go-Go Years" by writer Mark Reisner, the post-World War II era was the most active period of large dam construction in U.S. history. Glen Canyon Dam represents one of the last large western water storage projects, and no comparable water project has been built since in the Colorado River basin. It was also completed at a time when many U.S. citizens were beginning to express concerns about the environmental impacts of large storage dams (Billington and Jackson, 2006; Reisner, 1986; Worster, 1985).

Complementing the growth of the basin's water storage infrastructure, the 1920-1965 period also witnessed the creation of a complex legal structure governing allocations of river flow. Collectively, this legal framework is known as the Law of the River and it consists of interstate compacts, international agreements, water delivery contracts, and myriad other legal obligations. Milestones within this body of agreements, legislation, and court rulings (many of which were forged in the 1920-1965 period) are the 1922 Colorado River Compact, the 1928 Boulder Canyon Project Act, 1944 and 1973 international agreements with Mexico, the 1948 Upper Colorado River Basin Compact, the 1956 Colorado River Storage Project (CRSP) Act, the landmark Supreme Court decision (1963) and decree (1964) in *Arizona v. California*, the 1968 Colorado River Basin Project Act, the 1992 Grand Canyon Protection Act.

Colorado River Water Storage and Delivery Infrastructure

Hoover Dam and Lake Mead

Through the late 19th and early 20th centuries, the Colorado River's hydroelectric power potential attracted little attention because of a paucity of local demand. Similarly, the Colorado River was remote from any urban settlement that might seek to tap its flows (Figure 2-1). But by the 1920s economic conditions in the southwestern United States—especially in the greater Los Angeles region of Southern California—were changing rapidly, and earlier doubts about the economic viability of exploiting the river for municipal growth had largely disappeared. In this light, the driving force behind Boulder Dam can be traced to Southern California political and economic interests that were tied to both the Imperial Valley and to rapidly urbanizing Los Angeles. In addition, the U.S. Reclamation Service (renamed the Bureau of Reclamation in 1923) had long advocated the need for a dam on the lower Colorado in the name of comprehensive river development. After the 1905 flood that damaged existing water control infrastructure along the river in California, the

FIGURE 2-1 Colorado River upstream of the site of Boulder/Hoover Dam, ca. 1925. This stretch of the river today lies inundated by Lake Mead (impounded by Hoover Dam).
SOURCE: Postcard ca. 1925, no publisher.

need for greater river control became more pressing. The Bureau of Reclamation eventually merged its vision with the more immediate interests of Imperial Valley farmers and the City of Los Angeles for a significant flood control and water storage project on the river (Billington and Jackson, 2006; Hundley, 1992; Kleinsorge, 1941; Moeller, 1971).

Owing to a combination of geologic and climatic factors, the desert lands of southeastern California discharge only a minuscule amount of water into the Colorado River. Nevertheless, irrigators and civic boosters in Southern California were well positioned to lay claim to—and withdraw—huge quantities of water from the Colorado before any other states in the watershed could develop projects of comparable scale (Figure 2-2). By the early 1920s California legislators (with support from the Reclamation Service) were actively promoting federal construction of Boulder Dam. This, in turn, raised concerns among other Colorado basin states that feared completion of the dam would allow California to divert a large portion, perhaps most, of the river's flow. In addition, the huge storage project could

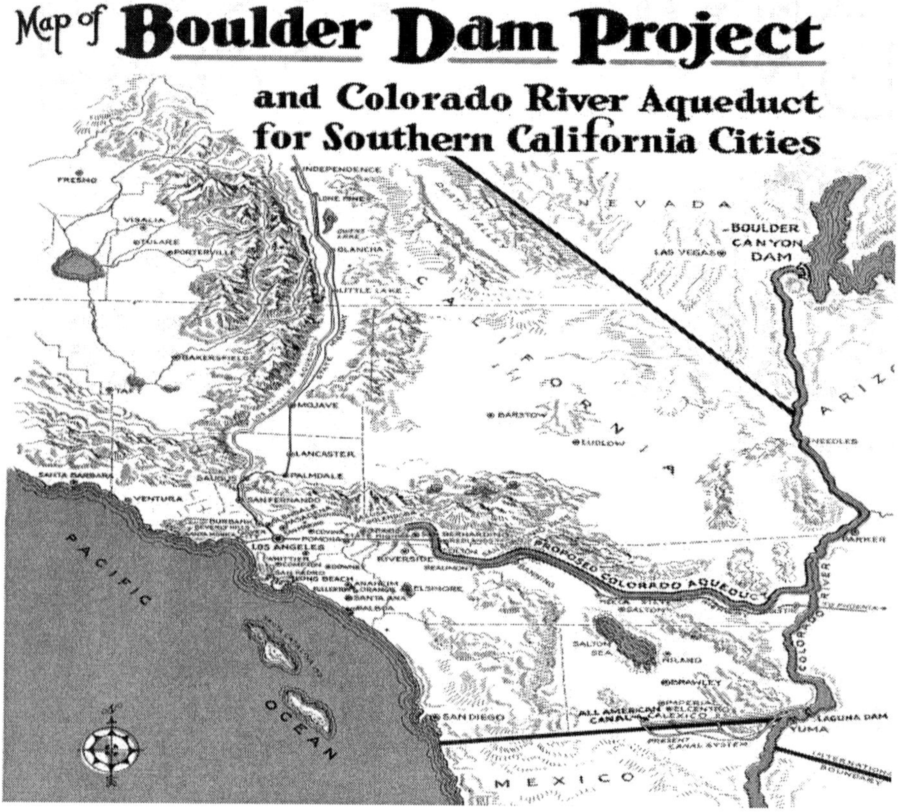

FIGURE 2-2 Map of proposed Boulder Dam Project and Colorado River Aqueduct, ca. 1930.
SOURCE: Metropolitan Water District of Southern California.

only be justified economically if financing was guaranteed by the sale of hydroelectric power, something that the privately financed electric power industry opposed. Taking all these factors into account, Senate approval of the Boulder Canyon Project Act required significant (although not necessarily unanimous) support from the same western states that feared California's monopolization of the Colorado River. The result of various sets of state and federal negotiations was the Colorado River Compact (described more fully below), a politically driven agreement between the upper basin and lower basin states dividing rights to Colorado River flows (Billington and Jackson, 2006; Brigham, 1998; Hundley, 1975, 1992; Kleinsorge, 1941; Moeller, 1971).

Following passage of the Boulder Canyon Project Act in 1928, construction of the 726-foot-high Boulder/Hoover Dam was started in 1931 under authority of the Bureau of Reclamation. Completed in 1935, Hoover Dam today impounds Lake Mead, a reservoir with a storage capacity of more than 28 million acre-feet[1] (Figure 2-3). By 1937 hydroelectric power from the dam was being transmitted to Southern California, and by 1940 power was used to pump water through the Metropolitan Water District of Southern California's Colorado River Aqueduct, a major water conduit serving domestic and industrial water supply needs in Los Angeles and surrounding cities (Bissell, 1939; Kleinsorge, 1941; Stevens, 1988).

FIGURE 2-3 Hoover Dam, ca. 1940.
SOURCE: U.S. Bureau of Reclamation.

---

[1] Twenty-eight million acre-feet is roughly enough water to supply the service area of the Metropolitan Water District of Southern California—which serves the coastal plain of Southern California from Ventura southward to San Diego—for 7 years.

Glen Canyon Dam and Lake Powell

Glen Canyon Dam was authorized as part of the 1956 CRSP Act, which authorized other water projects for the upper Colorado River basin (see following section). Following passage of the CRSP in April 1956, engineers and surveyors were at the dam site in July, and in October 1956 the first ceremonial blast was set off on the canyon wall (Rusho, unpublished manuscript). Construction of the dam was staged from the construction town of Page, Arizona, and, although a labor strike shut down construction of the dam for a short period in 1959, the dam and its power plant were completed on schedule (Figure 2-4 shows Glen Canyon Dam under construction). Construction of Glen Canyon Dam neared completion in 1963, at which time its diversion tunnels were closed and Lake Powell began to rise (and eventually was filled in 1980). Officially dedicated in 1966, Glen Canyon Dam stands over 700 feet high and impounds Lake Powell, which has a storage capacity of 27 million acre-feet. The dam is located 15 miles downstream from the Arizona-Utah border and 11 miles upstream from Lees Ferry. With a reservoir comparable in size to Lake Mead, storage provided by Lake Powell helps ensure that the upper basin states meet their water delivery obligations to the lower basin. Glen Canyon Dam feeds water into a large hydroelectric power plant, and power revenues were used to help finance construction costs (see Martin [1989] for more details on the construction of Glen Canyon Dam).

Colorado River Legal Framework: The Law of the River

The term "Law of the River" refers not to a single law, but rather to a complex array of agreements, legislation, court decisions and decrees, contracts, and regulatory schedules relating to the Colorado River, including a treaty with Mexico, two major multistate agreements (or compacts), Supreme Court rulings, and myriad other federal and state laws, acts, and regulations. In many ways, the foundation of the Law of the River was defined in the 1920s by the Colorado River Compact and the Boulder Canyon Project Act; over the ensuing decades, it expanded and evolved in numerous ways to incorporate new demands and shifting social and economic trends.

FIGURE 2-4 Glen Canyon Dam, nearing the end of its construction and prior to the beginning of water storage in 1963.
SOURCE: U.S. Bureau of Reclamation.

The Colorado River Compact (1922)

Signed in 1922 at Bishop's Lodge near Santa Fe, the Colorado River Compact is a cornerstone of the Law of the River. In terms of water law, a key impetus for negotiation of the Compact derived from the U.S. Supreme Court's decision in *Wyoming v. Colorado* (259 U.S. 419 [1922]). This ruling, which involved a dispute over use of the Laramie River, accorded the doctrine of prior appropriation interstate effect; that is, a diverter who appropriated water from an interstate stream in one state held a priority claim over a user in another state who diverted water from the stream at a later date. As the Supreme Court noted some 40 years later in the *Arizona v. California* case, this

1922 decision prompted concerns in the upper basin states that California's impending appropriation and use of Colorado River water stored behind the prospective Boulder Dam would result in California being "first in time" and therefore "first in right" with regard to later claims made by upper basin irrigators and cities.

At the Bishop's Lodge conference the seven basin states were unable to reach agreement on a state-by-state apportionment of Colorado River flows. The states instead divided the basin in half, designating Lees Ferry[2] on the Colorado River in northern Arizona (near the Arizona-Utah border) as the boundary point separating the upper and lower basins. Specifically, the Compact defines the upper basin to include much of Colorado, Utah, and Wyoming, and smaller parts of northern Arizona and northwestern New Mexico, whereas the lower basin consists of most of Arizona and portions of California, Nevada, New Mexico, and Utah (Figure 1-1).

As negotiated in 1922, the Compact apportions to both the upper basin and lower basin states the "exclusive beneficial consumptive use" of 7.5 million acre-feet annually.[3] To accommodate year-to-year variations in river flow, the Compact provides that the upper basin states will not deplete the flow at Lees Ferry by more than 75 million acre-feet for any consecutive 10-year period (Ingram et al., 1991). Put another way, the upper basin states agreed to provide an aggregate flow of at least 75 million acre-feet to the lower basin states (as measured at Lees Ferry) over any 10-year period. The Compact also provided that any water legally granted to Mexico at some future time would be shared equally between the upper and lower basins (Ingram

---

[2] Article II(e) of the 1922 Colorado River Compact defines "Lee Ferry" as "a point in the main stream of the Colorado River one mile below the mouth of the Paria River." Paragraphs (f) and (g) of Art. II define the terms "Upper Basin" and "Lower Basin," respectively, using Lee Ferry as the dividing point. Thus, in terms of water law, Lee Ferry is the appropriate legal term to describe the point at which the river's supply is divided between the Upper and Lower Basins. In the 1870s this stretch of the river came to be known as Lee's Ferry, after Mormon authorities directed John D. Lee to establish a crossing site for Mormon settlers heading south into Arizona. In 1921 the U.S. Geological Survey established a gaging station just upstream of the mouth of the Paria River. It was designated by the USGS as the Lees Ferry gaging station, and the station has retained this name for more than 80 years (Topping et al., 2003). As this report largely focuses on river hydrology and stream flow measurements, the term "Lees Ferry" is used to refer to this reach of the Colorado River.

[3] Additional flows of 1 million acre-feet per year from Colorado River tributary streams in the State of Arizona were later allocated to Arizona, pursuant to the 1963 *Arizona v. California* Supreme Court decision.

et al., 1991). The State of Arizona objected to the terms within the Compact, however, because Arizona did not want the flow of the Gila River (a tributary of the Colorado) to count against its allocation of Colorado River flows. Arizona therefore refused to ratify the 1922 Compact. The other states eventually approved the Compact and proceeded with a "Six State" Compact as provided for within the Boulder Canyon Project Act in 1928.

The Boulder Canyon Project Act (1928)

Key provisions of this federal legislation included (1) declaration that the Colorado River Compact would be effective upon the approval of six basin states if California, by state law, limited its guaranteed use to 4.4 million acre-feet of water per year; (2) authorization for building Boulder/Hoover Dam and the All-American Canal; and (3) an authorization to divide the lower basin share of 7.5 million acre-feet per year among the three lower basin states, with California being allocated 4.4 million acre-feet per year, Arizona receiving 2.8 million acre-feet per year, and sparsely populated Nevada receiving 300,000 acre-feet per year. The act also accorded the Secretary of the Interior broad authority over delivery of water stored behind Hoover Dam. Pursuant to this authority, the Secretary entered into contracts with Arizona and Nevada for the water allocated to them (even though neither state had physical means to divert and use the water) and with California water agencies for that state's shares, plus additional available unused water. The upper basin states did not conclude any agreement among themselves affirming the allocations stipulated in the Boulder Canyon Project Act.

Water Deliveries to Mexico (1944)

Signed in 1944, the *Treaty Between the United States of America and Mexico Respecting Utilization of Waters of the Colorado and Tijuana and of the Rio Grande* (59 Stat. 1219, T.S. 994) codified obligations of the United States to deliver water from the Colorado River to Mexico. During negotiations leading up to the 1944 treaty, Mexico proposed a division of the water that would acknowledge its right to increased flow as agricultural water demand in Mexico grew. The United States, which had already divided 15 million acre-feet of

Colorado River water per year between the upper and lower basins in the Colorado River Compact, balked at the prospect of signing a treaty with terms that could change in response to future use within Mexico. In the face of Mexico's proposition, the Colorado River basin states urged the United States to invoke the Harmon Doctrine of territorial sovereignty and assert the right to use every drop of Colorado River water flow within the United States without any obligation to deliver water to Mexico. The position associated with the Harmon Doctrine was not reflected in the treaty's final wording, however, primarily because Mexico insisted that conflicts over the Rio Grande (some of whose waters originate in Mexico) be negotiated at the same time (Meyers, 1967).

Reflecting a spirit of compromise, the 1944 treaty guarantees that the United States will deliver to Mexico the amount of 1.5 million acre-feet annually of the "waters of the Colorado River, from any and all sources." The treaty further provides that "Mexico shall acquire no right . . . for any purpose whatsoever, in excess of 1.5 million acre-feet of water annually," thus effectively blocking adjustments to the allocation based on international law or the use of "surplus" flow. Despite the language of guaranteed deliveries, the treaty contains provisions for relief in extreme circumstances, with guaranteed delivery of 1.5 million acre-feet per year subject to reduction in the event of shortages or drought upstream in the U.S. portion of the basin.[4] It does not provide specifically for water of a given quality, but this did not constitute a significant issue on the Colorado River until many years later (see the discussion later in this section on a 1973 Minute between Mexico and the United States that addresses the quality of water delivered at the Mexico-U.S. border).

---

[4] Three conditions must be present before deliveries to Mexico can be reduced: (1) "extraordinary drought" (a term not defined in the treaty) or some accident to the irrigation system; (2) "difficulty" to the United States in making deliveries of 1,500,000 acre-feet—without any provision as to who will determine that difficulty does, in fact, exist; and (3) reduction in U.S. consumptive uses (Meyers, 1967). The last condition is worth noting, as it requires that 1.5 million acre-feet be delivered annually to Mexico unless consumptive uses within the United States are decreased. The treaty does not specify the specific sources within the United States of Mexico's 1.5 million acre-feet annual allocation.

The Upper Colorado River Basin Compact (1948)

Signed in 1948, the Upper Colorado River Basin Compact apportions the 7.5 million acre-feet of consumptive use per year allocated to the upper basin under the 1922 Compact as follows: Colorado receives 51.75 percent, New Mexico receives 11.25 percent, Utah receives 23 percent, and Wyoming receives 14 percent. Arizona receives a fixed quantity of 50,000 acre-feet per year in recognition of its territory that drains into the river above Lees Ferry.[5]

The Colorado River Storage Project (1956)

After World War II the four upper basin states pushed for projects that would serve their interests; in 1956 Congress responded by authorizing the CRSP and its plans for developing several upper basin water storage projects. The key facility authorized within the CRSP is Glen Canyon Dam. In addition to Glen Canyon, the CRSP includes other major upper basin storage units, most notably Flaming Gorge Dam on the Green River in northeastern Utah; Navajo Dam on the San Juan River in New Mexico; and the multidam Wayne N. Aspinall Storage Unit on the Gunnison River in west-central Colorado (Martin, 1989; Sturgeon, 2002).

Arizona v. California (1963)

The *Arizona v. California* Supreme Court case settled a long-standing dispute over claims to Colorado River flow. In this landmark decision the Court issued both an opinion (373 U.S. 546 [1963]) and a decree (376 U.S. 340 [1964]). Arizona filed its original suit against California in the Supreme Court in 1952, and Nevada, New Mexico, Utah, and the United States were subsequently added as par-

---

[5] This allocation is a percentage of "beneficial consumptive use." Meyers reports that "[t]he Upper basin contends that the term means net depletion of the virgin flow; the Lower basin sometimes contends that it means consumptive use at the site of use, that is, the net loss to the stream at the place of use" (Meyers, 1967). The Upper Colorado River Commission, established by the Upper Colorado River Basin Compact, determines the sufficiency of supply within the upper basin to meet the delivery obligation at Lees Ferry and then determines the amount of any curtailment in the upper basin that is required to meet this obligation.

ties to the proceedings. As expressed by the Court, "[t]he basic controversy in the case is over how much water each State has a legal right to use out of the waters of the Colorado River and its tributaries" (373 U.S. 546 [1963]). Following the usual practice in such suits, the Court appointed a Special Master to take evidence, find facts, state conclusions of law, and recommend a decree. Appointed in 1955, Special Master Simon H. Rifkin held more than 2 years of formal hearings and, in 1961, submitted a 433-page report to the Supreme Court. This report contained the Special Master's findings, conclusions, and recommendations, most of which the Court adopted in its majority opinion and decree.

The decision in *Arizona v. California* established several important elements of the Law of the River. Finding that the dispute was controlled by the Boulder Canyon Project Act of 1928, the Court held that in passing the act, Congress created a comprehensive scheme for the apportionment of the lower basin's share of the mainstream waters of the Colorado River; it also reserved to Arizona, California, and Nevada the exclusive use of the waters of each state's own tributaries. This latter finding was of special importance to Arizona which, because of its "particularly strong interest in the Gila, intensely resented the Compact's inclusion of the Colorado River tributaries in its allocation scheme" (373 U.S. 546 [1963]) and, largely for that reason, was the only state that refused to ratify the Compact in the 1920s.[6]

The Court further concluded that the Boulder Canyon Project Act reflected a decision by Congress that a "fair division" of the first 7.5 million acre-feet of the Colorado's mainstream waters "would give 4,400,000 acre-feet to California, 2,800,000 to Arizona, and 300,000 to Nevada," and that "Arizona and California would each get one-half of any surplus." Moreover, the Court went on to hold that allocation of the water in these shares did not depend on the lower basin states agreeing to them in a compact because "Congress gave the Secretary of the Interior adequate authority to accomplish the division." It did so "by giving the Secretary power to make contracts for the delivery of water and by providing that no person could have water without a contract." Through these contracts the Secretary could not only implement allocations among the lower basin states, but could also decide which users within each state would get water. With reasoning that broke new ground in U.S. federal water law, the Court held that

---

[6] Arizona ratified the Colorado River Compact in 1944.

the Secretary was not bound by the law of prior appropriation in allocating water among the states nor by that doctrine or other priorities under state law in "choosing between users within each State" (373 U.S. 546 [1963]). Indeed, the Court found that the Act of necessity invested the Secretary of the Interior with sweeping powers over Colorado River management, especially in the lower basin:

> Today, the United States operates a whole network of useful projects up and down the river . . . . All this vast, interlocking machinery—a dozen major works delivering water according to congressionally fixed priorities for home, agricultural and industrial uses to people spread over thousands of square miles—could function efficiently only under unitary management, able to formulate and supervise a coordinated plan that could take account of the diverse, often conflicting interests of the people and communities of the Lower basin States. Recognizing this, Congress put the Secretary of the Interior in charge of these works and entrusted him with sufficient power . . . to direct, manage, and coordinate their operation (373 U.S. 546 [1963]).

Finally, the Court upheld claims of the United States to water in the mainstream of the Colorado River and in some of its tributaries for use on Indian reservations, national forests, recreational and wildlife areas, and other federal government lands and works. Specifically, the court reached a finding consistent with the *Winters* Doctrine of 1908, which established the principle of federal reserved water rights.[7] The Court upheld the decision reached in the *Winters* case, affirming that Indian water rights were created with, and dated back to, the establishment of a reservation(s). The rights of Indian tribes whose reservations predate passage of the Boulder Canyon Project Act thus are entitled priority. The Court also upheld another principle from the *Winters* case in finding that the United States intended to reserve sufficient water to satisfy not only present but also future needs of the Indian reservations; and that "enough water was reserved to irrigate all the practicably irrigable acreage on the reservations" (373 U.S. 546 [1963]).

---

[7] See *Winters v. United States*, 207 U.S. 564 (1908).

RELATIVE SURPLUS AND SHIFTING PRIORITIES: 1965 TO THE MID-1980s

The third phase of Colorado River water development and use extended from roughly 1965 through the mid-1980s. Although the pace of large water project construction decreased during these years, the huge increase in storage capacity added in 1920-1965 afforded generally ample water supplies in this post-1965 period. The water storage system proved sufficient to meet supply "shortfalls," which occurred primarily during droughts (e.g., during the late 1970s). But with population and economic growth, construction of fewer large water projects, and increasing concerns regarding instream flows for river ecology and recreation, the latter years of this era saw signs that existing supplies might not be able to deliver full benefits to all users.

With passage of the Endangered Species Act of 1973, endangered species became a concern that affected construction of proposed water development projects and operation of existing ones. Water quality and salinity levels also gained standing as significant and problematic issues. Environmental impacts resulting from the creation of reservoirs became manifest and the political viability of large-scale dam building waned. A notable event in this era occurred in the late 1960s, when plans by the U.S. Bureau of Reclamation to build hydroelectric power dams just upstream of Grand Canyon National Park (at Bridge and Marble canyons) were blocked in the U.S. Congress. Some big, new projects were initiated during this period—notably the Central Arizona Project (CAP)—but water project construction fell off the pace set during the 1950s and early 1960s.

Central Arizona Project (1968)

In concert with legal challenges to California's Colorado River claims, as early as the 1940s Arizona boosters sought federal support for a project to carry Colorado River water to central and southern Arizona. For many years the proposed CAP languished because of opposition from California. As indicated above, in its 1963 *Arizona v. California* decision, the U.S. Supreme Court found that in the Boulder Canyon Project Act Congress had given Arizona the right to withdraw 2.8 million acre-feet of flow annually from the mainstem of the Colorado River as part of a comprehensive scheme to apportion water among the lower basin states. With legal affirmation that the state's 2.8 million acre-foot annual allocation did not include flow originating within the Gila River (a Colorado River tributary), Arizona legislators undertook a push to win federal authorization of the CAP. As envisaged in the mid-1960s, the Central Arizona Project was to include hydroelectric power dams at Bridge and Marble canyons, but this aspect of the project aroused concern over potential impacts in Grand Canyon National Park (Figure 2-5).[8] In place of the hydroelectric power dams, Arizona legislators accepted a scheme in which power for pumping CAP water would come from a new coal-fired generating plant built on the Navajo reservation near the Utah border. And Arizona assuaged California legislators by agreeing to recognize California's claims to Colorado River flow as holding senior rights over Arizona's claims. Signed into law by President Johnson in 1968 as part of the Colorado River Basin Project Act, the CAP and its Granite Reef Aqueduct took decades to construct. When completed in 1992, the CAP was capable of delivering 1.5 million acre-feet of water per year—over half of Arizona's allocation as stipulated by the Boulder Canyon Project Act and affirmed by the U.S. Supreme Court—to the greater Phoenix and Tucson metro areas (Hundley, 1992; Pearson, 2002; Sturgeon, 2002).

Water Quality at the Mexico-U.S. Border: Minute 242 (1974)

When Mexico and the United States signed the 1944 treaty, water quality generally was not a significant issue. After 1944, with increasing population in the U.S. portion of the basin and as diversions for irrigated agriculture increased, Colorado River salinity became increasingly important. In the headwaters of the Colorado River, rain and snowmelt start out as essentially pure water. As the river and its tributaries flow downstream toward the Gulf of California, salts naturally accumulate as surface water and groundwater seep through subsurface salts and then into the stream channel. Colorado River salinity levels have risen in recent decades as low-salinity waters have

---

[8] The CAP plan also involved a preliminary proposal to divert water from the upper Snake/Columbia River watershed into the Colorado basin. By 1968 both the so-called "Grand Canyon Dams" and plans to divert Snake/Columbia water had been dropped from CAP legislation because of political resistance (Pearson, 2002).

FIGURE 2-5 Colorado River in Grand Canyon National Park, ca. 1940.
SOURCE: Postcard c. 1940, no publisher.

been diverted out of headwater areas for municipal and agricultural uses. Because many soils in the Colorado River basin have large amounts of naturally occurring salts, return flows from irrigated agriculture have also contributed to increased salinity levels. For example, in the 1980s return flows from Colorado's Grand Valley added an estimated 580,000 tons of salt each year to the Colorado River (Marston, 1987).

Of all the U.S. irrigation projects that affect Colorado River salinity levels, none are more important than the Wellton-Mohawk Irrigation and Drainage District near the mouth of the Gila River in southwestern Arizona. The salinity problem arose in 1961 with the District's discharge of drainage water into the Gila River. In 1962, the District began discharging drainage into the Bureau of Reclamation's Main Outlet Drain, which discharged water into the Colorado River. In 1961 Mexico protested that the highly saline water it was receiving from Wellton-Mohawk return flow was unsuitable for agricultural uses and that crop production in the Mexicali Valley was being adversely affected. In the early 1970s Mexico and the United States agreed upon a prospective and hoped-for solution to the salinity problem, and in 1973 the International Boundary and Water Commission

(IBWC) adopted Minute 242, bearing the formal title *Permanent and Definitive Solution to the International Problem of the Salinity of the Colorado River* (IBWC, 1973). This Minute 242 requires the United States to adopt measures to ensure that 1.36 million acre-feet of water delivered annually to Mexico upstream of Morelos Dam has an average salinity of no more than 115 ± 30 parts per million over the annual average salinity of Colorado River water arriving at Imperial Dam (*http://www.usbr.gov/dataweb/html/crwq.html#general*).

In 1974 Congress passed the Colorado River Basin Salinity Control Act, which authorized construction, operation, and maintenance of works in the Colorado River basin to control the salinity of water delivered to users in the United States and Mexico. Title I of the Act provided means for the United States to comply with its obligations under Minute 242; in addition, Title II created the Colorado River Basin Salinity Control Program and charged the U.S. Department of the Interior and the U.S. Department of Agriculture to manage the river's salinity, including salinity contributed from public lands (see *http://www.nrcs.usda.gov/PROGRAMS/salinity/*).

The Yuma Desalting Plant was constructed as a key facility to help comply with the 1974 Act and meet obligations set forth in Minute 242. Completed in 1992, the plant is the world's largest brackish water reverse osmosis desalting plant (*http://www.usbr.gov/lc/yuma/facilities/ydp/yao_ydp.html*). Prior to the plant's construction, a bypass drain was installed as an interim measure to divert saline drainage water from the Wellton-Mohawk irrigation project in Arizona away from the Colorado River mainstem and south to the Cienega de Santa Clara in Mexico. This was done in an effort to maintain acceptable levels of salinity of Colorado River water delivered to Mexico. The Yuma Desalting Plant was tested upon completion in 1992 but the facility was soon mothballed after heavy flooding along the Gila River in early 1993 destroyed parts of the canal that delivered water to the plant from the Wellton-Mohawk District. Since then, Colorado River salinity standards at the U.S.-Mexico border have been met through bypass of drainage water (which also represents a less expensive option) to the Cienega de Santa Clara.

Glen Canyon Environmental Studies (1982)

Glen Canyon Dam was constructed before enactment of the National Environmental Policy Act of 1969. As a result, a formal environmental impact statement (EIS) was not conducted as part of the planning studies for Glen Canyon Dam construction. In the early 1980s the Bureau of Reclamation sought to upgrade the hydroelectric power generators at Glen Canyon Dam and adjust operations to increase the dam's peak generating capacity. These changes could have had significant impacts on river flows, but it was not entirely clear from a legal perspective that an EIS would be necessary. Nevertheless, it was clear that the Bureau of Reclamation would have to assess, in some manner, the potential impacts of these changes at Glen Canyon Dam on the downstream riparian environment. As a result, in 1982 the Glen Canyon Environmental Studies (GCES) program was initiated to conduct this environmental research.

The GCES program was conducted in two phases—from 1982-1988 and from 1988-1996—and arrived at several findings relevant to Colorado River management (NRC, 1999). One finding from GCES was that Glen Canyon Dam and its operations have impacted the downstream environment and will continue to affect many ecosystem resources. GCES also demonstrated the value of an environmental monitoring system in managing ecological resources downstream of a large dam. GCES concluded that operation and management of Glen Canyon Dam could be modified to minimize losses of some resources, and to protect and enhance others. Through these findings, GCES provided a foundation for changes and adjustments to Glen Canyon Dam operations—including a highly publicized controlled flood released in March 1996 (Webb et al., 1999). GCES provided essential input into the Bureau of Reclamation's 1995 EIS on the operations of Glen Canyon Dam. It also laid the groundwork for a subsequent monitoring and scientific program—the Grand Canyon Monitoring and Research Center—that continues today.

TIGHTENING SUPPLIES AND INCREASING DEMANDS: MID-1980s TO THE PRESENT

The fourth phase of water development across the region began roughly in the mid-1980s and continues into the 21st century. This phase is characterized by limited development of new water supplies via traditional, structural means; rapid population growth and urbanization; an increasing emphasis on urban water efficiencies; and a shift of water supplies away from the agricultural sector to municipal and industrial users. This period has seen and continues to witness large population increases in the basin's major cities such as Las Vegas, Phoenix, and Tucson, as well as several cities on the basin's periphery that depend on Colorado River water, such as Albuquerque, Denver, Los Angeles, and San Diego. In some instances, increasing water demands caused by population increases have been partly offset by practices such as water pricing, new technologies, and conservation measures. But the overall effect of rapid regional population growth in this period has been to increase demand, causing municipalities to seek additional water sources. This has entailed numerous agricultural-urban water transfers across the region, such as a highly publicized transfer of water from the Imperial Irrigation District to the San Diego County Water Authority. In addition to serving as potentially valuable new supplies for municipalities, the trend of water shifting from agricultural to urban users has important economic, ecological, social, and cultural implications. This period has also seen a shift in the definition and vision of new water projects. In earlier periods, the vision of new water projects entailed dams, reservoirs, and conveyance facilities. Across the West today, new water projects are more likely to entail desalination plants, landscaping programs, water conserving technologies, underground storage, and educational programs designed to limit per capita uses (Chapter 4 reviews water conservation and augmentation efforts in the region).

Grand Canyon Protection Act (1992) and the Adaptive Management Program (1996)

In addition to limited water supplies and the growth of urban centers in the West, the contemporary era of Colorado River development features a water storage infrastructure that faces challenges of meeting traditional supply needs, along with relatively new demands—especially recreation and instream flows that support endangered species and distinctive ecological habitats. Two good examples of efforts in the Colorado River basin to balance shifting interests among a broad group of stakeholders are reflected in the Grand Canyon Protection Act and the Glen Canyon Dam Adaptive Management Program.

The Grand Canyon Protection Act of 1992 (P.L. 102-575) calls for the Secretary of the Interior to operate Glen Canyon Dam in accordance with the Law of the River and "in such a manner as to protect, mitigate adverse impact to, and improve the values for which Grand Canyon National Park and Glen Canyon National Recreation Area were established, including, but not limited to natural and cultural resources and visitor use." The act calls for the Secretary of the Interior to define operating criteria for Glen Canyon Dam in consultation with the Bureau of Reclamation, the Fish and Wildlife Service, the National Park Service, the Department of Energy, basin states, Indian tribes, and members of "the general public" that include environmental and recreational interests. It also called for completion of a final Glen Canyon Dam EIS; in response, the Bureau of Reclamation issued its Glen Canyon Dam EIS in 1995. This 1995 EIS and the 1992 Grand Canyon Protection Act serve as the primary guidance documents for the Adaptive Management Program.

The Glen Canyon Dam Adaptive Management Program was established in 1996 by the Secretary of the Interior to develop modifications to Glen Canyon Dam operations and to exercise other authorities under existing laws as provided in the Grand Canyon Protection Act. Its broad intent was to establish both a participatory stakeholder group and an ecological monitoring program, which were to help implement management decisions that would be studied and occasionally revisited, all of which would lead to more flexible, adaptive resources management (see Holling [1978], Lee [1999], and Walters [1986] for more on adaptive management; see Gloss et al. [2005] and NRC [1999] for more on adaptive management within the Grand Canyon ecosystem). The Adaptive Management Program was not intended to satisfy all mandates of the Grand Canyon Protection Act or to derogate any agency's resources management responsibilities. Rather, the Adaptive Management Program recommends administrative provisions, but these recommendations do not supersede basic management responsibilities of any of its cooperating entities (USBR, 1995). Adaptive Management Program constituents include the Secretary of the Interior's designee, an Adaptive Management Work Group, a Technical Work Group, and the Grand Canyon Monitoring and Research Center (see NRC [1999] and *http://www.usbr.gov/uc/rm/amp/index.html*).

Population Growth and Increasing Water Demands

A key driver affecting Colorado River basin water demands from the mid-1980s to the present has been rapid increases in population in many areas of the western United States served by Colorado River water. This population growth is being driven by a combination of migration from other U.S. states, immigration, and natural growth rate (birth rates minus death rates). Table 2-1 lists 1990-2000 population growth rates for several U.S. states and Figure 2-6 maps demographic changes across the entire United States for the same period. The sharp increase in growth rates in Arizona and Nevada, as well as in Colorado and Utah, is evident. In fact, these four Colorado River basin states were the fastest-growing (in percentage terms) U.S. states during the 1990s. From 1995 to 2005, population of the seven Colorado River basin states grew by nearly 11 million, an increase of roughly 25 percent (Griles, 2004).[9] These high percentage rates of population growth certainly stand out, but they should be considered along with absolute numbers of population growth. For example, the State of Nevada's 66 percent rate of population growth in the 1990s was the highest rate in the United States during this period. The State of California exhibited a far lower percentage rate of population

TABLE 2-1 U.S. Population Growth, 1990-2000

		Census Population		Population Change	
Rank	State	April 1, 2000	April 1, 1990	Number	Percent
1	Nevada	1,998,257	1,201,833	796,424	66.3
2	Arizona	5,130,632	3,665,228	1,465,404	40.0
3	Colorado	4,301,261	3,294,394	1,006,867	30.6
4	Utah	2,233,169	1,722,850	510,319	29.6
5	Idaho	1,293,953	1,006,749	287,204	28.5
6	Georgia	8,186,453	6,478,216	1,708,237	26.4
7	Florida	15,982,378	12,937,926	3,044,452	23.5
8	Texas	20,851,820	16,986,510	3,865,310	22.8
9	N. Carolina	8,049,313	6,628,637	1,420,676	21.4
10	Washington	5,894,121	4,866,692	1,027,429	21.1
12	New Mexico	1,819,046	1,515,069	303,977	20.1
18	California	33,871,648	29,760,021	4,111,627	13.8
32	Wyoming	493,782	453,588	40,194	8.9

SOURCE: http://www.census.gov/population/cen2000/phc-t2/tab03.xls.

---

[9] In the 1990s, no major U.S. metropolitan area grew faster in percentage terms than did Las Vegas, which grew at a remarkable rate of 83.3 percent.

growth during the 1990s; but during this period, while Nevada's population was increasing by 800,000 people, California added over 4 million people.

Studies and essays that consider the limits of western U.S. water resources and growth date back well over 100 years. Knowledge of "The Great American Desert," for example, was recorded on maps dating back to the report of Zebulon Pike of 1810 (Stegner, 1954). There have since been many debates over the limits and potentials for development of the region. These differing perspectives were prominently represented in the contrasting viewpoints of William Gilpin, first territorial governor of Colorado, and the western explorer and scientist John Wesley Powell. Whereas Gilpin saw nearly unlimited potential for western growth and settlement, Powell saw limits posed by the region's aridity that would require careful scientific management and a different approach to settlement than in the humid eastern United States (see Stegner, 1954).

Over the ensuing years of western growth and water development, there has been a paradigm that water supplies will always be available to satisfy ever-expanding population. And this paradigm was generally fulfilled as long as more rivers and groundwater supplies were available to be tapped. As western populations have grown, this historical approach has become less viable than in a previous era, leading to questions and debates about the limits of western growth and water supplies. These different views were pointed out in a proceedings of a 1991 National Academy of Sciences conference on western water and climate variability: "In all the major arid states, unlimited population growth is taken as an article of faith and the function of water policy is to supply all the water necessary to accommodate this growth. Many serious observers of the West think that the question is backwards. We should first set growth limits and use them to temper water demands to the more realistic use of available, possibly diminishing supplies" (Tarlock, 1991). Today's population levels and growth rates, intensifying competition for limited water supplies, and nontraditional water uses such as recreation and instream flows suggest that the relationships between population and urban water supplies soon will have to be confronted more seriously than in the past.

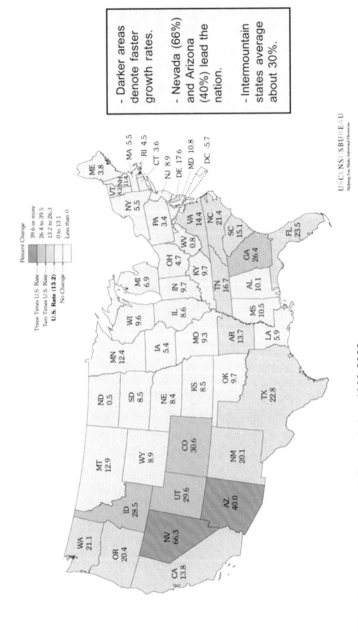

FIGURE 2-6 Percent change in U.S. population, 1990-2000.
SOURCE: *http://www.doi.gov/water2025/populate.html*.

Scientists studying climate and drought across the region are making connections between population growth, increasing water demands, and ability to cope with drought. A 2004 report on the western U.S. drought, for example, notes that

> The current drought is amplified by increased water demand in the southwest. This highlights the importance of evaluating all the possible causes of a decreased water supply. A mild hydrologic drought combined with the overuse of water supply can cause extreme drought condtions in a basin (Hidalgo, 2004).

It has also been noted that increasing demands on water supply will make it more difficult for the Colorado River storage system to recover from drought:

> Although six consecutive years of below-average inflow has not occurred in the past 100 years, we should not bet on a turn-around. More likely, storage will continue to decline in the near-term and the system will take longer to recover than it did after previous droughts—*largely because of greater demands today.* These increasing demands, including more use by the Upper basin states as they develop their allocated shares of the river, will mean less storage, on average, regardless of how long the current drought lasts (Fulp, 2005a, emphasis added).

Beyond increasing water demand, population growth across the region portends a range of negative impacts that include deteriorating air quality, additional urban "sprawl" and congestion, reductions in open space, and increased levels of traffic and strained transportation systems. Each additional person entails an increase in water demand of, roughly, at least 140 gallons per day (and often more).

Population growth in the West is contributing to increasing water demands. Population growth figures, however, do not necessarily equate as direct surrogates for regional water demand figures. Per capita water use is affected by several factors including water prices, household habits and preferences, landscaping choices, and public education programs. Many municipalities across the Colorado River region have implemented measures that have helped limit or reduce per capita water demands in many areas. Many federal, state, and local and municipal water organizations have generated water demand forecasts to help anticipate future per capita demands. The generation of accurate demand forecasts represents a data-intensive and analytical challenge, however, and there is a history of flawed demand forecasts at a variety of scales (see CRS [1980] and Rogers [1993] on challenges associated with water demand forecasting). With increasing population growth and a limited ability to extend water supplies through traditional means, accurate demand forecasts are as important as ever for water planning. As a scientific field, however, regional water demand forecasting lags behind much research on hydroclimatic and other water resources issues.

Increasing population growth rates and water demands in the 1990s and early 2000s have prompted many water users and managers to consider nontraditional means to extend water supplies. For example, groundwater sources have been increasingly tapped over the past few decades; water tables have dropped precipitously in many areas and the limits of groundwater resources are being approached in some areas (see Box 2-1). One prominent development on this front has been the sales, leases, and transfers of agricultural water to meet the needs of expanding urban populations.

Agriculture-Urban Water Transfers

Transfer of water from agricultural to municipal and industrial users in the U.S. West is not a new phenomenon. A well-known example of such a transfer was the purchase of agricultural water rights in rural Owens Valley by the City of Los Angeles (Gottlieb and Fitzsimmons, 1991; Kahrl, 1982). Today, these agricultural-urban transfers are taking place in many sites across the region including Colorado's South Platte River Basin (Denver), Las Vegas, and the Phoenix and Tucson metropolitan areas. In strict monetary terms, the sales, leases, and transfers of water from agricultural to urban users often represent "win-win" transactions for the buyer and seller, as water typically shifts from (in dollar terms) lower-value agricultural uses to (in dollar terms) higher-value urban uses. Municipalities and industries generally have a greater willingness to pay for a given unit of water than irrigators or ranchers, and these transfers may offer a cost-effective way for cities to meet increasing water demands. They also often prove profitable to individual farmers or ranchers.

> **Box 2-1**
> **Groundwater Resources in the Colorado River Region**
>
> Groundwater serves as an important water source for many people and municipalities across the nation. According to the U.S. Geological Survey, groundwater is the source of drinking water for about half the the nation and nearly all of its rural population (USGS, 2003). As in many regions of the United States, Colorado River basin groundwater resources have been heavily utilized to satisfy agricultural, municipal, and industrial water demands. High rates of groundwater pumping have led to depletion of aquifers in some areas, such as in southern Arizona. Use of groundwater to support population growth in south-central Arizona (primarily the Tucson and Phoenix metropolitan areas) has resulted in declines of water tables of between 300 and 500 feet (USGS, 2003). Rapid population growth in the Las Vegas region has also led to more groundwater pumping and declining water tables (up to 300 feet in some areas; USGS, 2003). High rates of groundwater extractions have also led to lowered water tables in peripheral areas along the Colorado River basin, in both Southern California and New Mexico.
>
> Groundwater is a valuable resource in the arid West and its pumping provides a wealth of social and economic benefits. Intensive pumping of groundwater, however, may entail negative impacts. One important concern related to groundwater extraction is reduced flows from groundwater systems to streams. Surface and groundwater systems are usually linked, and groundwater extraction may alter how water moves between aquifers and streams. Lowered water tables can inhibit groundwater flow into streams, or increase the rate at which water moves from a surface body into an aquifer. In either case the impact is a reduction of flows to surface water, which can lead to the

The amount of water devoted to agricultural uses across the West is not insignificant. The 75-80 percent of western U.S. water supplies presently diverted to agriculture represents many millions of acre-feet of water. In the State of Arizona, for example, water allocated to the agricultural sector as of 2006 exceeded 5 million acre-feet per year (Table 2-2). Table 2-2 illustrates several important points. One is a striking rate of population growth and water demand: the period 1990-2040 is forecast to experience roughly a doubling in municipal and industrial water demand (roughly 2 percent annually). Another point within this table is the large percentage of water currently devoted to agricultural uses—nearly 80 percent as of 1990. This leads

> loss of riparian vegetation and wildlife habitat. A prominent regional example of the impacts of groundwater extraction on riparian ecology is in the Santa Cruz River near Tucson. The Santa Cruz River valley once supported a large assemblage of mesquite, cottonwood, and other species, and provided important wildlife habitat. Over time, water extractions to meet the demands of a growing population have caused declining water tables, leading to reduced surface water availability and a large loss of riparian vegetation (USGS, 2003).
>
> In response to high rates of groundwater extraction in many areas of the state, the State of Arizona has enacted legislation that represents some of the nation's most stringent guidelines surrounding groundwater use. In 1973, the Arizona legislature enacted the Adequate Water Supply Program, a law requiring land developers to obtain a statement of water adequacy from the (former) Arizona Water Commission (Davis, 2006). In 1980, the Arizona legislature enacted a Groundwater Management Act to conserve groundwater resources. These areas were legally defined as Active Management Areas and they are centered on the state's largest urban and agricultural centers (they do not cover the entire state). Passage of the Groundwater Management Act saw the Adequate Water Supply Program replaced by the Assured Water Supply Program (Davis, 2006). Perhaps the most notable administrative difference in the new program is that if a subdivider within an Active Management Area fails to demonstrate an assured water supply to the Arizona Department of Water Resources, the Arizona Department of Real Estate cannot approve the subdivision for sale, the county cannot record the plat, and the developer thereby is prevented from selling lots. Whether these new regulations will help to noticeably resolve groundwater pumping issues in the State of Arizona remains to be seen; but state officials and water providers clearly take problems related to groundwater overdraft seriously and are seeking measures to remedy them.

to another observation: modest portions of water reallocated from this large amount of water in agricultural uses to the municipal and industrial sector can help satisfy increasing municipal and industrial demands.

Modest shifts (in percentage terms) of agricultural water to municipal and industrial uses, therefore, can do much to quench increasing urban water demands. Although this water can serve as a valuable supply for growing urban areas, these shifts are not without costs and limitations. There are direct effects associated with water rights being transferred out of agriculture, such as reduced food

TABLE 2-2 Water Demands in Arizona

Category	Water Demand (acre-feet)			A-F Change 1990-2040	% Change 1990-2040
	1990	2015	2040		
Municipal and industrial	1,332,000	1,922,000	2,605,000	1,273,000	96
Agriculture	5,339,000	5,220,000	5,037,000	- 302,000	- 6
STATE TOTAL	6,671,000	7,142,000	7,642,000	971,000	15

SOURCE: http://geochange.er.usgs.gov/sw/impacts/society/water_demand/.

Historical and Contemporary Aspects of Colorado River Development

production capability. Another important consideration is that such changes in points of diversion and water uses nearly always entail "third-party" effects beyond those that accrue to the buyer and seller of water rights. Examples of these effects include reduced agricultural return flows that support riparian ecosystems, and lost business suffered by local merchants as a result of reductions in irrigated cropland. The various costs that may be borne by such third parties are well recognized (e.g., NRC, 1992). If not addressed carefully and equitably, effects on third parties can be a cause of conflict. For example, water being transferred from a rural area to a municipality may negatively affect agriculture-related businesses (e.g., farm machinery dealers) that depend on irrigated agriculture, which may be harmful to small western U.S. farming communities (Howe et al., 1990).

Another factor that may inhibit transfers is limited physical infrastructure, especially water storage and conveyance facilities, available to facilitate transfers. Several innovative and useful practices—especially water banking and aquifer storage—have been developed to help obviate the need for new storage and conveyance facilities (see Box 2-2 for a discussion of the Quantification Settlement Agreement, a comprehensive and prominent arrangement for transferring water from agriculture to urban users). Nevertheless, there will always be some physical limitations on potential transfers across the basin. Many creative water transfer programs, involving legally defined water banks and underground water storage programs, have been developed to help effect these transfers, but the amount of agricultural water is finite and such programs thus are necessarily limited in their ability to satisfy ever-increasing demands over the long term. Certain basin states also have consistently opposed leasing, trading, or selling of water beyond state boundaries, although many complex, recent agreements are designed to allow flexibility while protecting allocations.

The waters diverted by the agricultural sector likely represent the final large source of water that municipalities in the Colorado River region will be able to draw upon to significantly support urban growth. As this finite "source" of water approaches its limits in being transferred to municipalities and industries, urban users will be increasingly pressed to adopt more stringent conservation and regula-

> **BOX 2-2**
> **Moving Water from Agriculture to the Cities:**
> **California's Quantification Settlement Agreement**
>
> For many years, the State of California diverted more than the 4.4 million acre-feet of Colorado River allocated within the Boulder Canyon Project Act because a portion of the lower basin's allocation of 7.5 million acre-feet remained unused by the other lower basin states of Arizona and Nevada. With increasing population and increasing water demand for Colorado River water in these other states, however, it became essential for California to limit its annual diversion to 4.4 million acre-feet. The key to California's meeting its commitment was an agreement among its southern farming and urban communities on the way in which its share would be allocated. This accord took the form of the 2003 Colorado River Water Delivery Agreement: Federal Quantification Settlement Agreement among the Secretary of the Interior, Imperial Irrigation District, Coachella Valley Water District, Metropolitan Water District of Southern California, and the San Diego County Water Authority (or the CRWDA).
>
> The CRWDA was signed on October 16, 2003, at Hoover Dam by the Secretary of the Interior and four Southern California water agencies. Its cornerstone is an agreement by the Imperial Irrigation District, California's largest user of Colorado River water, to transfer up to 200,000 acre-feet annually to the San Diego County Water Authority for up to 75 years. This long-term agriculture-to-urban water transfer is the largest in U.S. history and will supply San Diego with about one-third of its future water needs.
>
> Under a set of Interim Surplus Guidelines agreed upon by the seven basin states and the Department of the Interior in 2000, the relevant California water agencies were to adopt a Quantification Settlement Agreement (QSA) by December 31, 2002. This QSA represents an agreement among Imperial Irrigation Disrict, Coachella Valley Water District, and the Metropolitan Water District of Southern California. If California met this and other benchmarks, it would continue to have access to more than its share of Colorado River water during a transition period, making possible a so-called "soft landing" as it gradually reduced its use to the allocation of 4.4 million acre-feet as specified in the 1928 Boulder Canyon Project Act. However, if it missed a benchmark it would immediately lose access to all water above that amount, resulting in a "hard landing." The latter outcome followed from the failure of the rural and urban agencies to adopt a QSA by the end of 2002. The sharp reduction in the water supply spurred new negotiations which, though difficult, finally produced the 2003 QSA. This agreement led the Secretary of the Interior to reinstate the "soft landing" features of the Interim Surplus Guidelines.

tory measures in order to stretch existing supplies (Chapter 4 of this report discusses technological and other prospects for augmenting water supplies). Water transfers will no doubt continue to be used to meet increasing water demands, and municipalities, tribes, farmers, and other water users will continue to develop innovative means for effecting these transfers. But growing populations will nonetheless act to offset "gains" in water supplies achieved by transfers. It is recognized that population growth and higher water demands are reducing the region's ability to cope with drought and increasing the potential for conflicts over limited water supplies (see, for example, the Department of Interior's Water 2025 website: *http://www.doi.gov/water2025/*). The limits of Colorado River water supplies, increasing populations and water demands, warmer temperatures, and the specter of recurrent droughts point to a future in which tension and conflict among existing and prospective new users are likely to be endemic.

Challenges of meeting water demands always increase during periods of drought. As described in the following section, the early 2000s saw below-normal precipitation across much of the Colorado River basin, which resulted in sharp decreases in inflows into Colorado River system reservoirs.

Early 21st Century Drought

A severe, multiyear drought across much of the western and southwestern United States in the early 21st century had substantial impacts on Colorado River basin water supplies. Figure 2-7, for example, illustrates changing patterns of precipitation and the worsening drought across the region from 2000 to 2006. The nature of drought makes it difficult to identify exact dates on which it may have begun and ended (see Box 2-3). By one measure—inflows into Lake Powell—drought conditions existed from 2000 to 2004 (Table 2-3). By other measures, drought conditions extended beyond 2004 and affected portions of the basin in late 2006 (see Piechota et al. [2004] for an evaluation of 1999-2004 drought conditions across the basin).

Reduced amounts of precipitation and inflows resulted in substantial drops in reservoir storage levels in the late 1990s and early 2000s. In 1999, reservoirs on the Colorado River were more than 90 percent full, but by 2005 system-wide storage had fallen to about 50 percent—a decrease in volume of some 25 million acre-feet of water

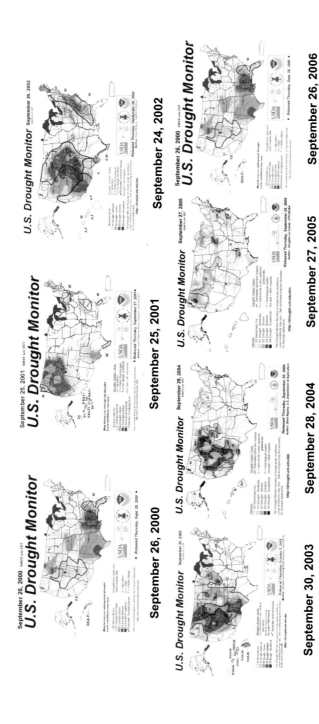

FIGURE 2-7 Western U.S. drought conditions, 2000-2006.
SOURCE: *http://www.drought.unl.edu/dm/archive.html*. The U.S. Drought Monitor is a partnership between the National Drought Mitigation Center (NDMC), U. S. Department of Agriculture, and National Oceanic and Atmospheric Administration. Map courtesy of NDMC-UNL.

BOX 2-3
Defining Drought

Clear definitions of drought are elusive. Drought is generally understood in terms of the definition offered in Webster's Dictionary: dryness; want of rain; or a prolonged period of dryness. Drought is a normal part of climate in nearly all of the United States but it is of special concern in arid regions of the western United States, where precipitation is often in short supply and where one thus might say drought exists much of the time.

Drought can be defined in different terms, including meteorological, agricultural, hydrologic, and socioeconomic (Wilhite and Glantz, 1985). Hydrologic definitions of drought are of particular interest within this report, as Colorado River water managers generally define drought in terms of reservoir inflows. The Colorado River basin drought of the early 21st century saw well below normal inflows into Lake Powell for the 5-year period 2000-2004. It should be noted, however, that 1999 and 2005 both had only slightly above-normal inflows, and one or two years of slightly above normal inflows do not end a drought of such magnitude. For 1999-2005, average inflows into Lake Powell were below normal. The 2006 water year is likely to extend this trend.

A basic concept invoked in understanding drought is that of a water budget. Water is held in storage buffers such as soil root zones, aquifers, lakes, reservoirs, and surface stream flows. These buffers act as water supplies, are subject to demands, and are replenished and lose water at varying rates. When losses exceed replenishment, impacts are experienced and, at lower storage levels, become increasingly severe. In essence, drought is defined by its impacts on both natural and man-made environments because without impacts there is no drought, no matter how dry it might be. Drought infers a relationship between supply rates and demand rates; drought is not simply a supply-side phenomenon, but also depends on water demands. Without demands, there is no drought, whether a given supply of water is big, small, or even zero.

It can be difficult to determine exactly when a drought has begun or ended, and there can be differences of opinion over whether a drought actually exists. Droughts begin slowly. They may be interrupted by wet periods, during which it is not clear if precipitation will continue or if dry conditions will return. A drought may not be widely recognized until it has been under way for several months or longer, and it can be particularly difficult to recognize in arid regions that experience seasonal dry periods. Recognizing that drought began in some parts of the Colorado River basin in the late 1990s, and that it is ongoing in many areas and may not abate any time soon, this report uses the descriptors of *drought of the early 21st century* and *drought of the early 2000s* to refer to the drought that has affected Colorado River hydrology in this period.

TABLE 2-3 Unregulated Inflow to Lake Powell

Water Year[a]	Percent of Average
1999	109%
2000	62%
2001	59%
2002	25%
2003	52%
2004	51%
2005	109%

[a] "Water year" refers to the period from October 1 to September 30 of the following year (e.g., Water Year 2004 refers to the period from October 1, 2003, until September 30, 2004).
SOURCE: Fulp (2005b).

(Fulp, 2005a). In early 2005 Lake Powell was at its lowest level of storage since 1969, when it was initially filling (Figures 2-8 and 2-9), and Lake Mead had not been as low since 1967 (Fulp, 2005a). The drought of the early 2000s was severe by any measure; in terms of climate statistics, the probability is very low—less than 0.1—that any 5-year drought period since 1850 had been as dry as 2000-2004 (Woodhouse et al., 2006).

During the early 21st century drought, the Colorado River storage system performed much like it had been designed to do and, even after 5 consecutive below-average years of precipitation and inflows, still held roughly 2 years of annual Colorado River flows (Fulp, 2005a). Precipitation across the Colorado River basin was closer to average conditions in 2005, but in 2006 drier conditions returned and were exacerbated by above-normal temperatures; July 2006, for example, was the second-warmest-ever month of July in the continental United States (http://www.noaanews.noaa.gov/stories2006/s2677.htm), and the 2006 average annual temperature for the contiguous United States was the warmest on record (and nearly identical to the record set in 1998; http://www.ncdc.noaa.gov/oa/climate/research/2006/ann/us-summary.html).

It is not clear how drought will impact future reservoir storage levels, nor is it clear exactly how long it would require the storage system to refill once again. According to one estimate, it may require roughly 15 years of average hydrology to refill Lakes Powell and Mead (Jeanine Jones, California Department of Water Resources, personal communication, 2005). Closer-to-normal precipitation may

Historical and Contemporary Aspects of Colorado River Development   65

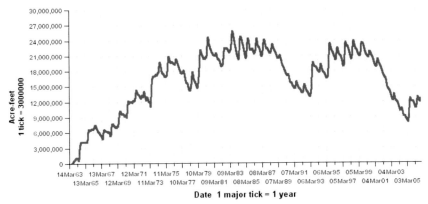

FIGURE 2-8 Storage in Lake Powell through December 1, 2006. Lake Powell's capacity is 27 million acre-feet, including dead storage. Values shown do not include the volume of water in dead storage.
SOURCE: Generated at *http://www.usbr.gov/uc/crsp/GetSiteInfo*.

FIGURE 2-9 Glen Canyon Dam and Lake Powell, August 2004. Note the residual ring around the top of the lake caused by declining water levels.
SOURCE: Courtesy of Brad Udall, University of Colorado.

end the drought, but a return of drier conditions could extend it. Regardless of future precipitation conditions, higher levels of population and water demand will make it more difficult to fill reservoirs and meet future water demands and obligations.

Coping with Drought and Increasing Water Demands

The early 21st century drought has been notable for its hydrologic and related impacts, such as forest fires in some areas of the Colorado River basin. The drought, along with increasing population growth and water demands, stimulated a variety of responses. It is not possible to list here every new water use and drought mitigation strategy across the West, but this section introduces some notable drought responses in the early 2000s (Chapter 5 includes more detail on drought mitigation programs and studies).

In response to drought conditions and increasing competition over the West's water resources, the Department of the Interior initiated a program to help increase awareness of possible future conflicts over water, especially during drought. Entitled *Water 2025: Preventing Crises and Conflict in the West*, the program was started in 2003 in an effort to concentrate "existing federal financial and technical resources in key western watersheds and in critical research and development, such as water conservation and desalinization, that will help to predict, prevent, and alleviate water supply conflicts" (DOI, 2003). The Water 2025 program has provided limited funds for competitive "challenge grants," much of which have gone to agricultural conservation projects. It has also sponsored workshops across the western United States that convened scientists, engineers, and water managers to discuss water shortage problems and possible solutions. A key premise driving the Water 2025 initiative is that "In some areas of the West, existing water supplies are, or will be, inadequate to meet the demands for water for people, cities, farms and the environment even under normal water supply conditions" (*http://www.doi.gov/water 2025/Water2025-Exec.htm*). Water 2025 produced a map (see Figure 2-10) of areas across the western United States that may experience water supply crises by the year 2025. This figure shows that several areas of "highly likely" to experience conflicts lie within or adjacent to the Colorado River basin.

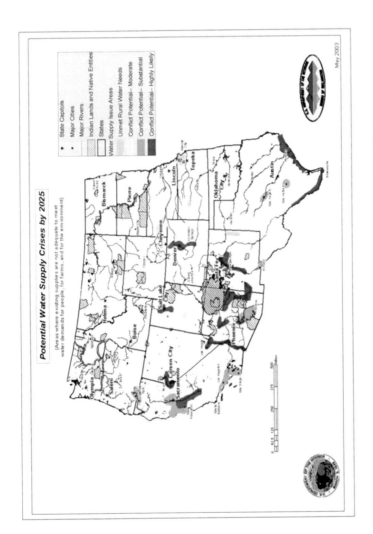

FIGURE 2-10 Potential water supply crisis areas in the western United States.
SOURCE: *http://www.doi.gov/water2025/supply.html*.

Colorado River basin states are engaged in a variety of long-range water planning, drought management, and conservation plans and programs. The State of Colorado, for example, in 2004 began a Statewide Water Supply Initiative that examined all aspects of the state's water uses through 2030 and discussed water supply options and management alternatives. The report's first finding was that "[s]ignificant increases in Colorado's population—together with agricultural water needs and an increased focus on recreational and environmental uses—will intensify competition for water" (CWCB, 2004). Another example of state-level planning for future water demands and shortages is the Arizona Drought Preparedness Plan. Issued in 2004 by the Governor's Drought Task Force, this report provides guidance to water users within Arizona and serves as a foundation for a long-term, statewide water conservation strategy. In 2006, the California Department of Water Resources issued an extensive report on incorporating climate change into California state water management (California DWR, 2006). The other basin states are also involved in plans and studies aimed at enhancing water conservation, drought planning, and long-term water supply availability (Chapter 5 includes further discussion of drought management programs and initiatives in the region).

An important development that grew out of drought conditions in the early 2000s was a letter of agreement signed by representatives of all seven Colorado River basin states (see Appendix A). Dated February 3, 2006, this letter was sent to the Secretary of the Interior in response to the Secretary's request for the states to develop shortage guidelines and management strategies under low-reservoir conditions. No basin-wide shortage criteria existed prior to the 2000s, and the Secretary had declared that the Department of the Interior would develop these guidelines if the basin states were unable to arrive at a consensus agreement. The letter and the level of cooperation it represents constitute an important step toward devising the first formal set of shortage criteria among the seven basin states and, as such, provide some optimism regarding future interstate cooperation on Colorado River water supply issues.

COMMENTARY

The past 150 years have been marked by four broad phases of Colorado River water development. The first era extended from the middle of the 19th century until roughly 1920. This period was characterized by explorations led by John Wesley Powell and by the origins of contemporary practices of irrigated agriculture. The second era extended from roughly 1920 to 1965. This period saw the signing of several crucially important water development agreements and rulings, and the construction of many major, multipurpose water projects. This era began with the planning for Boulder (Hoover) Dam and ended with the construction of Glen Canyon Dam. A third phase of Colorado River water development extended from 1965 until the mid-1980s. This period was characterized by ample water supplies that supported population growth and economic growth across a variety of sectors, including both urban and agricultural uses. Fewer dams and water projects were constructed during this period as compared to the 1920-1965 era. Water storage facilities constructed between 1920 and 1965 generally provided adequate water to support additional people and economic development. This provided a water supply "cushion" during periodic droughts, such as during drought across much of the basin in the late 1970s. The period also witnessed rising concerns regarding environmental impacts of large-scale dams and associated water supply systems.

The fourth phase of regional water development began in the mid-1980s and continues today. This phase is characterized by limited water supply development and rapid population growth and urbanization. During the 1990s the four fastest-growing states (in percentage terms) in the nation were Nevada, Arizona, Colorado, and Utah, respectively. The basin's major cities, such as Las Vegas, Phoenix, and Tucson all experienced large increases in population, as did several cities on the basin's periphery that depend on Colorado River water, such as Albuquerque, Denver, Los Angeles, and San Diego. Not only do these larger numbers of people increase urban water demands, many of these urban dwellers enjoy and support other, nontraditional uses of western water, namely instream flows for both recreation and environmental preservation. In some instances, population increases have been partly offset through water pricing and conservation measures that have reduced per capita demands, and by transfers of water from agricultural users. Increasing water demands are also encouraging a reordering of priorities to favor uses with a stronger economic base and users possessing a greater willingness to pay for water. But as population and water demand continue to grow, urban water supply gains realized by conservation, water transfers, and other measures are eventually absorbed. The impact of high, steady population growth has been to increase water demands; in the face of limited water supplies, these increasing and broadening demands portend a decreasing ability to cope with drought conditions and heightened conflicts over limited water supplies.

When the Colorado River Compact was signed in 1922, except for municipal water demands in Southern California, use of water for irrigated agriculture was a predominant concern. The basin was lightly populated and the river's water was allocated equally between the upper and lower basins. At that time, population in the lower basin states was roughly double that in the upper basin states. Since that 1922 allocation the upper and lower basins experienced different levels of population growth and urbanization. The most important demographic feature in the ensuing years was population growth in Southern California (and to a lesser extent in Arizona). Today, population in the lower basin states is four to five times the population in the upper basin states. Fueled by this growing population, the lower basin states eventually began to use their full 7.5 million acre-feet annual allocation of Colorado River water. By contrast, the upper basin states have never used their full allocation of 7.5 million acre-feet of water per year.

Releases of water from Glen Canyon Dam have always exceeded the upper basin's delivery obligation of not less than 75 million acre-feet for any 10 consecutive years, pursuant to Article III(d) of the Colorado River Compact. Even during the drought of the early 2000s and lowered water storage in Lake Powell, Glen Canyon Dam was delivering flows above the upper basin's Colorado River Compact commitment. There is no imminent prospect that this delivery obligation will not be met, and any change in the Colorado River Compact would require the resolution of numerous complex legal issues that could require many years or even decades to resolve. Nevertheless, the upper basin states intend to utilize a greater portion of their 7.5 million acre-feet per year allocation and, with rapid population growth in many areas, they continue to come closer to their full Colorado River Compact allocation. Future droughts and climate change may also affect precipitation and inflows into Lake Powell and other storage facilities. Any shortages in water delivery obligations that resulted from climate change would be dealt with in the same way as shortages caused by drought or other factors. If changes in climate rendered the Law of the River inadequate to deal with resulting shortages, the Colorado River basin states could conceivably seek to amend the Compact and the United States and Mexico could conceivably seek an amendment of their 1944 treaty. Water releases from Glen Canyon Dam are a key issue at the hydrology-climate-population growth nexus in the Colorado River basin and bear close watching in the years ahead.

One initiative that grew from the drought conditions of the late 1990s and early 2000s was the Department of the Interior's Water 2025 program. The following excerpt comments on the Water 2025 program and touches on the water supply and demand issues discussed in this chapter:

> On the very first page of its 2003 report, "Water 2025," the United States Bureau of Reclamation explains with chilling frankness that "today, in some areas of the West, existing water supplies are, or will be, inadequate to meet the demands of people, cities, farms, and the environment even under normal water supply conditions." The report goes on to explain "the reality" that: "explosive population growth in western urban areas, the emerging need for water for environmental and recreational uses, and the national importance of the domestic production of food and fiber from western farms and ranches are driving major conflicts between these competing uses of water. The 'major conflicts' are occurring because most all of the surface waters in the region have been appropriated, leaving little for the continuing stream of newcomers" (Gavrell, 2005).

Another critical issue is the prospect for transferring water from agricultural uses to help meet growing urban demands. Roughly 80 percent of the Colorado River basin's water is allocated to the agricultural sector. Given limitations on constructing (and filling) new storage reservoirs, growing western cities are looking to agricultural water as a source of additional supplies. Municipalities often have a large willingness to pay for agricultural water rights, and both parties (i.e., agricultural sellers and urban buyers) often stand to gain by these types of transfers. There is a large amount of water held in agricultural water rights that could support a great deal of future urban population growth. There are barriers to transferring water to municipalities from agricultural and other users, such as tribal groups. These barriers include direct and third-party effects, and limited physical facilities for storing and rerouting water among willing buyers and sellers. As agricultural supplies are diverted to urban uses, this last remaining substantial amount of water that could be made available for urban uses in the Colorado River region is, slowly but surely, being depleted.

Steadily rising population and urban water demands in the Colorado River region will inevitably result in increasingly costly, controversial, and unavoidable trade-off choices to be made by water managers, politicians, and their constituents. These increasing demands are also impeding the region's ability to cope with droughts and water shortages.

The drought of the early 2000s brought climate-related concerns to the fore across the Colorado River region. Not only did the drought result in numerous, direct hydrologic impacts, it raised questions about what climate trends and future conditions across the region and the planet might portend for Colorado River flows. The early 21st century also saw a great interest in several climate and hydrologic studies of the Colorado River region, especially several long-term reconstructions of past Colorado River flows that were based on studies of the annual growth rings of coniferous trees. The following chapter discusses how features of the global climate system affect the Colorado River region, temperature and precipitation trends and projections across the region, the gaged record of Colorado River flows, and studies of annual growth rings of coniferous trees (dendrochronology) and what they imply for regional hydrology and climate.

3

Climate and Hydrology of the Colorado River Basin Region

The Colorado River basin contains climate zones ranging from alpine to desert and exhibits significant climate variability on a variety of time scales. These variations have important implications for snowmelt and river hydrology and are thus of interest to both scientists and water managers in the Colorado River region. Scientific research on the Colorado River basin's climate and hydrologic systems has included measurements of the river's flow, long-term studies of climate and river hydrology, reviews of statistics associated with temperature and precipitation extremes, and studies of connections to regional and global climate systems. In the 20th century, long-term water management and planning in the region generally relied upon the gaged record of Colorado River flows; specifically, great reliance was placed on measurements made at Lees Ferry, supplemented by data recorded at other stations on the mainstem and on tributary streams. Some of these gaged streamflow records for the Colorado River date back to the late 19th century, but most began during the 20th century.

Although a time frame of over 100 years may appear to offer an extensive record of climate and streamflow variability, in fact it represents a relatively short period in terms of geologic history of the region. In recent years, the once-prevailing view of climate as static and unchanging on time scales important to river managers has given way to a new understanding that the gaged record represents only a small temporal window of the variability characteristics encompassing many centuries of Colorado River hydroclimate. River management decisions are inherently forward looking and rely heavily on forecasts. These forecasts typically assume that past properties of the river system, as revealed through observations, will be replicated in future conditions. However, the prospect of changing states of atmospheric conditions and climate behavior, associated with anthropogenic emissions of greenhouse gases, calls this assumption into question. As a result, many water managers today are exploring ways of adjusting water planning and management strategies.

The study of climates that occurred before direct measurements of weather and climate data—*paleoclimatology*—can serve as part of the hydroclimatic information considered in water management decisions. This field of study draws upon indirect, or proxy, information about past climate conditions obtained from evidence contained in glacial ice, landscape features, sediment deposits in ancient lakes, pollen, species distributions, preserved organisms (e.g., mollusks), and middens. The science of *dendrochronology*, or the study of the sequences of annual growth layers (rings) of coniferous trees, is particularly relevant in the Colorado River basin. For several decades, cores from coniferous trees in the western United States have been analyzed to enhance understanding of past climate. Recent tree-ring analyses have incorporated updated chronologies and longer calibration periods to estimate annual Colorado River flows over the past several centuries. These new dendrochronological reconstructions have stimulated heightened interest in questions regarding the rarity and recurrence of drought conditions across the region.

This chapter discusses fundamental features and dynamics of Colorado River basin climate (including climate trends and future climate scenarios), the gaged record of Colorado River streamflow, and tree-ring studies of past Colorado River region streamflow. The concluding Commentary section discusses implications of this hydrologic and climatic information for water resources planning and decisions.

FEATURES AND DYNAMICS OF COLORADO RIVER BASIN CLIMATE

Precipitation Patterns and Sources

The Colorado River is primarily a snowmelt-driven system, with most precipitation in the basin falling as winter snowfall in higher elevations of Colorado, Utah, and Wyoming. In the upper Colorado River basin, approximately 20 percent of the basin's precipitation falls in the highest 10 percent of the basin, and roughly 40 percent of the basin's precipitation falls in the highest 20 percent of the basin. Cold temperatures at high elevation cause precipitation to occur mainly as snow and to remain frozen during the winter months. This "white reservoir" drapes the mountain terrain during winter months and survives into summer at the highest locations. Some of the water in this snowpack is lost to the atmosphere through sublimation (a phase change from solid to vapor) during the cool season. Most remains, however, and as the snowpack warms, or "ripens," in the spring, meltwater is steadily metered into the soil. This process extends for several weeks to months at higher elevations, and melting occurs slowly enough to recharge the soil and allow water to enter the myriad channels that feed the Green and Colorado rivers. For these reasons, winter precipitation over the high-elevation portion of the upper basin plays an important large role in generating runoff and streamflow.

Warm season precipitation plays a different role in the basin's hydrology. During warmer months precipitation falls more intensely, often in localized, convective thunderstorms. Plants are photosynthetically active at all elevations and utilize some of this water immediately. Furthermore, almost all the summer precipitation intercepted by vegetation canopies evaporates directly to the atmosphere. Much of the remainder of summer precipitation that infiltrates into the soil column is transpired by plants or (in the case of bare ground) evaporates, aided by warm soil. A relatively small fraction of summer precipitation makes its way into aquifers and streams. In the basin's high-elevation headwaters, summer precipitation amounts are generally less than winter values. The high-elevation winter dominance of annual precipitation is more pronounced in the Green River drainage than in the Colorado River headwaters in central Colorado. In the basin's lower and drier reaches, summer precipitation can account for a larger share of annual total precipitation, but because of higher evaporation and transpiration rates, this moisture is less effective in contributing to streamflow. In the hottest and lowest portions of the basin, summer precipitation matters greatly to local vegetation and to small runoff channels, but hardly at all to the mainstem Colorado and its major tributaries.

The main source of summer moisture is the North American monsoon, which transports moisture into the region from sources in the subtropical Pacific and Gulf of Mexico. This annual phenomenon brings drama to the southwestern desert skies, but only occasionally does it provide enough precipitation to contribute appreciably to hydrologic supplies. For the mainstem Colorado River and its major tributaries, the bulk of the precipitation that contributes to water supply falls during the winter months, primarily in the form of snows at high elevation. Summer months comprise the period of higher water demands and, except in extreme weather years, will provide at best only modest additions to mainstem reservoir water supplies. If a season of winter precipitation and water storage is "lost" because of drought conditions, there will be little opportunity to replenish supplies until the following winter.

The Tropical Pacific and ENSO

Ocean temperature patterns that have the greatest influence on Colorado River basin climate are in the tropical Pacific in a band that straddles the equator between Peru and the International Date Line. At irregular intervals of typically 2-7 years, sea surface temperatures (SSTs) in this region warm above climatological averages.[1] This phenomenon, called El Niño, is part of a complex ocean-atmosphere oscillation. El Niño has a climatic counterpart called La Niña that is characterized by below-average SSTs (La Niña events usually have smaller departures from average SST than do El Niño events). The terms El Niño and La Niña refer only and exclusively to ocean temperatures in this geographic domain and not to their effects elsewhere.

Another atmospheric feature relates to barometric pressure gradients in the South Pacific. In the 1920s, British meteorologist Sir Gilbert Walker published his seminal work describing the inverse relationship in atmospheric surface pressure between Tahiti and Easter Island in the tropical Pacific, and over Darwin in northern Australia (Walker, 1925). That is, when atmospheric pressure is high in one of these locations it tends to be low in the other region, and vice versa. Walker termed this phenomenon the Southern Oscillation. It refers only to the atmosphere. The Darwin-Tahiti pressure difference (nor-

---

[1] Tropical Pacific SSTs are 1-3°C above average in modest El Niño events, 3-5°C above average in major episodes.

malized for variability over the past century) is the basis of the Southern Oscillation Index (SOI). Furthermore, when Tahiti has lower than average pressure and Darwin has higher than average pressure (negative SOI), a strong tendency exists for El Niño to be present. Conversely, there is a tendency for La Niña conditions to exist with higher pressure in Tahiti and lower pressure in Darwin. The oceanic (SST) and atmospheric (SOI) measures are usually highly correlated and these terms are sometimes used interchangeably (McCabe and Dettinger, 1999). For historical reasons these phenomena are often lumped together and referred to (although somewhat asymmetrically) as El Niño-Southern Oscillation, or ENSO. The ENSO phenomenon owes its existence to coupled ocean-atmosphere interactions over the equatorial Pacific and is an important contributor to interannual global climate variability. The ENSO cycle has impacts on climate over large areas of both the tropics and extratropics. Jerome Namias was the first to investigate extensively the possible relationship between SST and North American atmospheric circulation. Jacob Bjerknes identified the equatorial Pacific as the source of climate variability associated with the Southern Oscillation.

The winter storm track over the eastern Pacific Ocean shifts southward during El Niño episodes, often causing wet winters in the southwestern United States and dry winters in the Pacific Northwest and northern Rockies. La Niña winters tend to bring the opposite pattern, and moderately positive values of the SOI in the prior summer/autumn nearly guarantee a dry winter in the southwestern United States—it is the most dependable predictive climate relationship in the United States (Redmond and Koch, 1991). In Arizona and New Mexico, and extending into the San Juan Mountains of southwestern Colorado, El Niño winters are generally wetter than normal, but not always, and a few are extremely dry. Moreover, the likelihood of an extremely wet winter is much higher during El Niño winters and there are few wet winters when El Niño conditions are not present (Redmond and Koch, 1991). These patterns are accentuated in streamflow, particularly in extreme high and low streamflow (Cayan et al., 1999). Precipitation patterns in the western United States vary considerably among different El Niño events. These differences appear to depend on the particular spatial pattern of warm ocean temperatures, the magnitude of warming, and the particular months of the year when these patterns occur. Accurate forecasting of these ocean features and their North American effects represents one of today's principal ENSO-related forecasting challenges.

Within the Colorado River basin, ENSO effects are more pronounced in the lower basin than in the upper basin. The San Juan River shows the same strong relationship to ENSO as does Arizona. By contrast, the headwaters of the Green River (in Wyoming's Wind River Mountains) tend to be slightly more influenced by the northern pole (centered over the Columbia River basin) of this winter dipole pattern (Redmond and Koch, 1991). The main source regions of Colorado River basin precipitation and streamflow—the mountains of Colorado, Wyoming, and northeastern Utah—are not greatly impacted by ENSO events. Because roughly 90 percent of the river's flows originate in mountain headwater regions with limited connection to ENSO, better forecasts of ENSO and its effects are not likely to greatly improve upper basin mainstem streamflow forecasts.

Other Ocean Connections

Another pattern of Pacific regional scale climate variability related to SST variations is the Pacific Decadal Oscillation (PDO). The term was coined in 1996 by fisheries scientist Steven Hare while he was studying connections between the Alaska salmon production cycle and Pacific climate (*http://jisao.washington/edu/pdo*). The PDO describes joint variations in SST, atmospheric pressure, and wind in the central and eastern Pacific poleward of 20°N (Mantua et al., 1997). The warm and cool phases of the PDO each historically have lasted two to three decades, for a total period of about a half-century. An abrupt jump in Pacific-wide environmental conditions known as the "1976 shift" (Ebbesmeyer et al., 1991; Trenberth and Hurrell, 1994) was identified retrospectively and helped lead to identification of the PDO. This pattern appears to alternately accentuate and counteract the effects of ENSO in the Pacific Northwest and the southwestern United States and is expressed most strongly in winter. The origin of this oscillation has not been definitively determined. It is linked to periods of greater and lesser frequency of El Niño and La Niña at equatorial latitudes, even though the PDO index has only a modest correlation with the SOI (Mantua et al., 1997). Although there are intriguing statistical relationships associated with the PDO, the physical mechanisms that underlie the PDO behavior itself, and that lead to its expression within the Colorado River basin (and primarily in the lower basin, as is the case with ENSO), have not been fully explained.

In recent years another pattern has been identified that appears to have ties to the Colorado River basin. Atlantic Ocean SSTs exhibit a mode of variability that has similar departures from average for one to two decades over an area spanning low to high latitudes; this feature is known as the Atlantic Multidecadal Oscillation (AMO). That the AMO has effects on climate and streamflow in the eastern United States (Sutton and Hodson, 2005) is understandable; however, additional studies have shown some surprising results. Notably, when the North Atlantic is warm for a decade or longer, streamflow in the upper Colorado River basin tends to be lower than average, and vice versa (Gray et al., 2004; McCabe and Palecki, 2006; McCabe et al., 2004). This headwaters streamflow is largely governed by winter precipitation. The physical mechanism by which the Atlantic could influence mountain winter precipitation in Colorado and Wyoming, which are upstream in the atmospheric winter flow pattern, remains a puzzle. The evidence so far is statistical and largely dependent on just a few AMO cycles. Theory and models are just beginning to address this potential link (Delworth and Mann, 2000; Knight et al., 2005) and observational studies are continuing. For example, during warm Atlantic phases, moisture delivery to the conterminous United States is diminished (Schubert et al., 2004a).

Diagnostic studies of the global pattern of ENSO cycle variability clearly have revealed that the atmosphere acts as a bridge linking SST anomalies in the equatorial Pacific to yet larger patterns of atmospheric and ocean variability. Variations in SSTs in the tropical Pacific may herald changes in jet stream patterns, strength and track of Pacific winter storms, and future water supply conditions across the Colorado River basin. Different patterns may accentuate or counteract each other. For example, the effect of Indian Ocean temperatures acting in concert with La Niña has been demonstrated as helping produce "the perfect ocean for drought" in the southwestern United States (Hoerling and Kumar, 2003). Research has shown that the American Dust Bowl of the 1930s was in part caused by tropical ocean temperature departures from normal (Schubert et al., 2004b; Seager, 2006; Seager et al., 2005). Other western droughts, such as the droughts during the Civil War era and in the 1890s, may have similar explanations (Seager, 2006; Seager et al., 2005). Linkages among these patterns suggest modest predictability, enough that they may merit consideration in water supply planning across the western United States.

CLIMATE TRENDS AND PROJECTIONS

Climate Records and Past Trends

In a previous era of Colorado River water management there was an implicit assumption that the main features of future climate states would closely resemble those of the past century. Over time, additional research has enhanced understanding of the variability of past climate over longer time scales. Moreover, increasing levels of atmospheric greenhouse gases and steadily increasing global mean surface temperatures have heightened awareness of the potential for human activities to impact the global climate system (Houghton, 2004). The assumption that future climate conditions will largely replicate past conditions is now frequently being called into question.

Variations in precipitation and water supply have long been of interest to water managers for daily, monthly, and annual operations. Less widely appreciated are the impacts that temperature has on water availability, through effects on both supply and demand. Temperature affects the quantity of and timing of snowmelt runoff in spring and summer, the occurrence of large floods, and rates of evapotranspiration. Anything that affects basin temperatures in a long-term, systematic way thus should be of considerable interest, regardless of its origin. The observed time series of basin-averaged precipitation and temperature are important for assessing regional impacts of global climate change and are discussed in the following section.

Precipitation

Colorado River basin precipitation exhibits high year-to-year (interannual) variability. Figure 3-1 shows interannual precipitation variability across the upper Colorado River basin, spatially averaged over the basin upstream of Lees Ferry and aggregated to annual resolution (Kittel et al., 1997; updated data from *ftp://ftp.ncdc.noaa.gov/*

pub/data/prism100). For example, after a period of less variability for several decades in the mid-20th century, there has been a tendency toward greater variability in the latter decades of the 20th century. The past 30 years of data include the highest and lowest annual precipitation in the 100-year record, and there has been a tendency toward multiyear episodes of both wet and dry conditions. Some years in the early and mid-1980s were at least as wet as the period that preceded the signing of the Colorado River Compact. Prior to the early 21st century drought, the driest comparable 5-year consecutive interval was the 1950s drought. The only other comparable 5-year dry period was at the end of the 19th and beginning of the 20th century. Despite these variations, there is no significant trend in interannual variability of precipitation over the past 110 years.

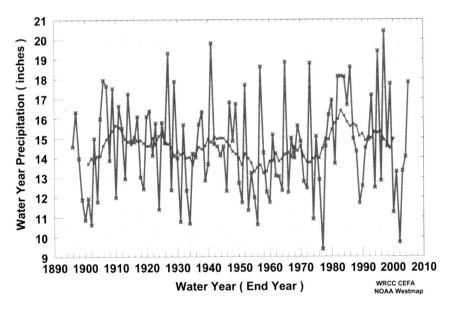

FIGURE 3-1 Annual precipitation for the Colorado River basin above Lees Ferry, 1895-2005.
NOTE: Red: annual values. Blue: 11-year running mean.
SOURCE: Western Regional Climate Center.

Temperature

Figure 3-2 shows annual mean temperatures for the entire Colorado River basin from 1895 to 2000 (adjusted for variations in elevation using the same method as for precipitation in Figure 3-1). Upper and lower basin temperature trends are similar and bear a strong resemblance to the history of temperature across the entire western United States (Redmond, in press), as well as to mean global surface temperature trends. Figure 3-2 shows that since the late 1970s the Colorado River region has exhibited a steady upward trend in surface temperatures. The most recent 11-year average exceeds any previous values in the over 100 years of instrumental records.

One striking aspect of Figure 3-2 is how much warmer the region has been in the drought of the early 2000s as compared to previous droughts. For example, temperatures across the basin today are at least 1.5°F warmer than during the 1950s drought. Increasing

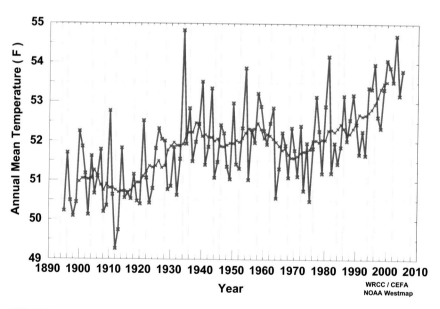

FIGURE 3-2 Annual average surface air temperature for entire Colorado River basin, 1895-2005.
NOTE: Red: annual values. Blue: 11-year running mean.
SOURCE: Western Regional Climate Center.

Climate and Hydrology of the Colorado River Basin Region

temperatures in the region have many important hydrologic implications, including the impacts of drought. For example, the drought of the early 2000s has taken place in particularly warm conditions. Figure 3-3 shows temperature departures for that 6-year period (2000-2005) as compared to 1895-2000 averages. Both in terms of absolute degrees and in terms of annual standard deviation, the Colorado River basin has warmed more than any region of the United States—a fact that should be of great interest throughout the region. This trend continued through the first half of 2006. This warming is well grounded in measured climatic data, corroborated by independent data sets, and widely recognized by climate scientists throughout the West.

The trend of increasing temperatures in the western United States also is seen in larger, global temperature trends. For example, a 2005 paper on western mountain snowpack trends notes that "increases in temperature over the West are consistent with rising greenhouse gases, and will almost certainly continue" (Mote et al., 2005). And in a recent review of global surface temperature records of the past 2,000 years, a committee of the National Research Council (NRC, 2006) concluded that

> It can be said with a high degree of confidence that global mean surface temperature was higher during the last few decades of the 20th century than during any comparable period during the preceding four centuries. This statement is justified by the consistency of the evidence from a wide variety of geographically diverse proxies (NRC, 2006).

Key manifestations of warmer temperatures in western North America are a shift in the peak seasonal runoff (driven by snowmelt) to earlier in the year, increased evaporation, and correspondingly less runoff. In fact, many of these changes have been documented:

> Winter and spring temperatures have increased in western North America during the twentieth century (e.g., Folland et al. 2001) and there is ample evidence that this widespread warming has produced changes in hydrology and plants. . . . The timing of spring snowmelt-driven streamflow has shifted earlier in the year (Cayan et al. 2001; Regonda et al. 2004 [*corr* Regonda et al., 2005]; Stewart et al. 2005), as is expected in a warmer climate (Hamlet and Lettenmaier 1999a) (Mote et al., 2005).

84

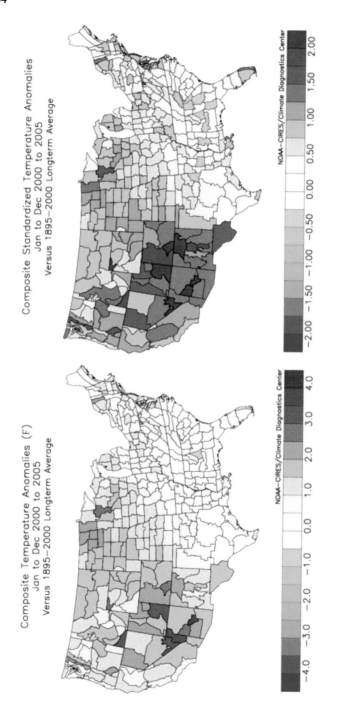

FIGURE 3-3 2000 to 2005 temperature departures from 1895-2000 average.
Left: Shown in temperature units (°F). Right: Shown in standardized terms (standard deviations).
SOURCE: National Oceanic and Atmospheric Administration's Climate Diagnostics Center.

A recent study of the timing of snowmelt in several mountain basins in the western United States concluded that "[t]he recent midlatitude warming, perhaps of anthropogenic origins, is a plausible cause for the shift in spring peak flow timing" (Regonda et al., 2005). Other studies of snowpack over the western United States find that declining trends in snow accumulation likely are not solely a manifestation of precipitation and snowfall variability, but rather reflect (at least in part) a warming signal:

> Estimates of future warming rates for the West are in the range of 2°—5° C over the next century, whereas projected changes in precipitation are inconsistent as to sign and the average changes are near zero (Cubash et al. 2001). It is therefore likely that the losses in snowpack observed to date will continue and even accelerate (Hamlet and Lettenmaier 1999a; Payne et al. 2004)   (Mote et al., 2005).

Projecting Future Climate Conditions

Many studies of future climate and hydrology conditions across the western United States are based on results of computer-based, numerical models of the global atmosphere. Developed in large part to project future effects of human-induced climate change arising from increasing levels of heat-trapping greenhouse gases, these atmospheric models—referred to as general circulation models (GCMs)—are used for a variety of experiments. These numerical models of the global climate system are the primary method used by climate scientists to project global and regional atmospheric responses to a variety of perturbations, such as a doubling of atmospheric carbon dioxide levels. Seasonal to interannual weather forecasts from multiple models are generally viewed to be more accurate than individual forecasts (Krishnamurti et al., 1999). This perspective regarding "consensus" weather forecasts can be generalized to climate forecasts, and has led to a trend of using multiple GCM output scenarios to assess implications of climate variability and change.

Precipitation Projections

For reasons similar to the difficulties in making daily precipitation forecasts, long-term projections of precipitation constitute a greater modeling challenge than temperature projections. Over the West and the Colorado River basin, precipitation projections from climate models suggest a wide range of potential changes in annual precipitation. Results from multiple computer runs, over many model-scenario combinations, generally forecast precipitation futures that show relatively little annual change in the region (see Dettinger [2005] for precipitation projections that are representative of the western United States). Over the next 10-40 years, there is a tendency in the results of climate model superensembles to forecast slightly increased annual precipitation in the northwestern United States by about 10 percent above current values, and to forecast slightly decreased annual precipitation in the southwestern United States by less than 10 percent below current values, with relatively little change in annual precipitation amounts forecast for the headwaters regions of the Colorado River.

Changes in seasonality of precipitation or changes in the type of precipitation (rain or snow) can be just as important as changes in annual amounts of precipitation. A detailed study for the Sierra Nevada mountains (at the same latitude as the upper Colorado), using 11 climate models and 2 emission scenarios, projects slightly more precipitation in winter and slightly less precipitation in late winter and in spring and early summer (Maurer, 2007). To a first approximation, no appreciable trend in annual Colorado River basin precipitation has been detected (Figure 3-1) or currently is projected. The accuracy of climate model precipitation forecasts is a topic of great interest and will continue to be an important focal point in climate science research.

Temperature Projections

Figure 3-4 compares multiple climate model projection results for temperature across the Colorado River region. Key points from these projections are the unanimity among the different models that temperatures will rise in the future, and relatively small differences across projections during the first part of the 21st century. Differences among these model results are modest until roughly 2030, with increasing divergences among them moving toward the year 2100. All these changes and model results are so far broadly consistent with recorded temperature data for the region. Taken as a whole, these future projections and past trends point to a strong likelihood of warmer future climate across the Colorado River basin.

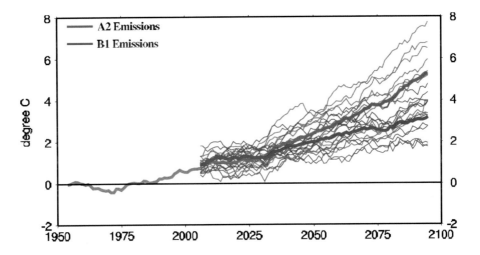

FIGURE 3-4 Nine-year moving average of observed annual air temperature averaged over the Colorado River basin (1950-2001), and projected Colorado River basin annual average air temperature from 11 different climate models, under two different greenhouse gas emission scenarios (2005-2095). Greenhouse gas scenarios were run for the Intergovernmental Panel on Climate Change (IPCC) Fourth Assessment Report (AR4).
Red (A2) projections are 9-year moving averages based on relatively unconstrained growth in emissions over the next century; solid red line represents a 9-year moving average of A2 projections.
Blue (B1) projections are 9-year moving averages based on a stabilization of global emissions by 2100; solid blue line represents a 9-year moving average of B1 projections.
SOURCE: Gridded observation data from Maurer et al. (2002). IPCC AR4 climate projections from Lawrence Livermore National Laboratory Program for Climate Model Diagnosis and Intercomparison (*http://www-pcmdi.llnl.gov*).

Hydrologic Implications of Warming

These projected temperature increases across the Colorado River region have important direct and indirect implications for hydrology and streamflow, irrespective of precipitation increases or decreases. The likely effects of warmer temperatures across the Colorado River basin for hydrology include the following:

- freezing levels at higher elevations, which means more winter precipitation will fall as rain rather than snow;
- shorter seasons of snow accumulation at a given elevation;
- less snowpack accumulation compared to the present;
- earlier melting of snowpack;
- decreased base flows from groundwater during late summer, and lowered water availability during the important late-summer growing season;
- more runoff and flood peaks during the winter months;
- longer growing seasons;
- reductions in soil moisture availability in summer and increases in the spring and winter;
- increased water demands by plants; and
- greater losses of water to evapotranspiration.

Concerns regarding the implications of future climate changes—especially warming—for Colorado River flows date back to at least the 1970s. Since then, the effects of the listed factors on Colorado River streamflow have been incorporated in different ways by several different hydrologic studies and papers. In a study often acknowledged as the first to evaluate possible impacts of climate change on Colorado River flows, Stockton and Boggess (1979) estimated that a 2°C increase in temperature and a 10 percent decrease in precipitation would result in a decline of upper basin streamflow of about 44 percent. In a 1983 paper, Revelle and Waggoner estimated that a 2°C temperature increase by itself would cause a decrease in mean Colorado River flows by 29 percent. Subsequent studies have used more sophisticated approaches based on hydrologic models that represent the physical processes that relate climate and streamflow, and generally have estimated somewhat less severe impacts on runoff resulting from prospective temperature increases (e.g., Nash, 1991). In the early 1990s, for example, a series of hydroclimate modeling studies indicated that hypothetical temperature increases of 2° and 4°C, and no change in precipitation, would lead to Colorado River streamflow reductions of 4-12, and 9-21 percent, respectively (Nash and Gleick, 1991, 1993).

A 2000 assessment of the potential consequences of climate variability and change on U.S. water resources considered the implications of changes in climate on runoff in the Colorado River basin (Gleick, 2000). Modeling exercises specially conducted for the assessment were based on output from two GCMs; these results included forecast increases of 66-128 percent in upper Colorado River flows (from that report's Table 7). In addition to these specific modeling exercises, the 2000 assessment lists results from several other hydroclimate modeling experiments and professional papers. In contrast to modeling results for the assessment that projected increases in Colorado River runoff, the majority of the results from these other hydroclimate modeling exercises project future decreases in runoff for the upper Colorado River and inflows into Lake Powell (see Table 9 from that report). In its review of these other modeling experiments and papers, the report notes that, "In the arid and semi-arid western United States . . . [e]ven in the absence of changes in precipitation patterns, higher temperatures resulting from increased greenhouse gas concentrations lead to higher evaporation rates, reductions in streamflow, and increased frequency of droughts." It was also observed that, for climate-runoff projections for several river basins in the semiarid western United States, "[i]n every one of these studies, an increase in temperature and no change in precipitation resulted in decreases in runoff" (Gleick, 2000).

A more recent study of the global consequences of 21st century climate change used average values from 12 different GCMs to project future runoff changes (Milly et al., 2005). Almost all the model runs projected future decreases in runoff over the interior western United States, including the Colorado River region. These decreases are projected to be on the order of 20 percent (Milly et al., 2005). Another study of western North America arrives at similar conclusions: reductions in annual runoff resulting from increasing temperature and slight decreases in precipitation (by 1-6 percent) may reduce Colorado River inflow to Lake Powell by 14-18 percent over the next half-century (Christensen et al., 2004).

Differences among these forecasts of future streamflow can be ascribed to modeling and other methodological differences. Some of these studies (e.g., Milly et al., 2005) are based on output from GCMs with relatively coarse resolution (typically 2-4° latitude-longitude) of the Earth's surface and atmosphere, which cannot resolve details of the relatively small areas from which most of the Colorado River's flow is generated. Aspects of the processes that generate runoff—such as negative feedback between earlier runoff and reduced evaporative demand associated with warmer winter temperatures in headwaters regions—thus are not adequately captured. Differences in GCM results can contribute to differences in hydrologic projections. For example, results in the 2000 U.S. National Assessment (Gleick, 2000) that projected future increases in runoff in the upper Colorado basin were based on the U.K. Hadley Centre model, which tends to simulate large precipitation increases relative to other GCMs. Other GCM-based projections suggest changes in seasonality of precipitation; a parallel climate model (PCM) from the National Center for Atmospheric Research forecasts little change in annual precipitation but shifts some winter precipitation to the summer months. In the Colorado River basin, summer precipitation is on average less efficient in generating runoff (because of higher evaporative losses) than in winter. As a result, runoff changes were amplified from the modest PCM warming projections (Christensen et al., 2004). For these reasons, recent studies have begun to utilize multimodel ensemble approaches and to focus on the ensemble mean, with the range of results used as an index of uncertainty. This approach was used in the Milly et al. (2005) global study and in a recent report of the California Governor's Climate Action Team (California Environmental Protection Agency, 2006). An ensemble-based approach to hydrologic and climate forecasting is becoming more widely applied and accepted.

A 2006 paper employed 11 different climate change models that are being used in the Intergovernmental Panel on Climate Change (IPCC) Fourth Assessment Report (AR4), which is due to be released in 2007 (Christensen and Lettenmaier, 2006). GCM output was used for two global emissions scenarios: an "A2" (relatively unconstrained growth in emissions over the next century) and a "B1" (elimination of global emissions increases by 2100) scenario. Results showed that, in the ensemble mean, Colorado River discharge at Imperial Dam (naturalized flow) would decrease by up to 11 percent by the end of the century for A2 emissions, and by up to 8 percent for B1 emissions. Over all ensembles, 9 of 11 showed streamflow decreases by the end of the century for A2, and 8 of 11 for B1—roughly the same fraction as in the results from the Milly et al. (2005) paper. In comparison with an earlier paper (Christensen et al., 2004), part of the reason noted for the somewhat smaller streamflow reductions predicted is that most of the IPCC AR4 scenarios show shifts (albeit modest) of summer to winter precipitation, which tend to counteract increased evaporative demand associated with warmer temperatures (Christensen and Lettenmaier, 2006).

There have also been some (but fewer) studies evaluating the implications of streamflow changes on reservoir system performance. A 1993 paper used a U.S. Bureau of Reclamation Colorado River reservoir simulation model with historic streamflows altered according to a range of precipitation and temperature changes (Nash and Gleick, 1993). They assumed an instantaneous temperature (or precipitation) change, so the results in this modeling exercise refer to the eventual equilibrium response. That paper found that a 20 percent reduction in Colorado River natural runoff would result in mean annual reductions in storage of 60-70 percent, reductions in power generation of 60 percent, and an increase in salinity of 15-20 percent at the U.S.-Mexico border (Nash and Gleick, 1993). A 2004 paper concluded that changes of up to 18 percent in runoff could result in somewhat smaller decreases—up to 40 percent—in total basin storage (Christensen et al., 2004). That study's authors noted, within the various climate and hydrologic scenarios they used, that "[r]eleases from Glen Canyon Dam to the Lower Basin (mandated by the Colorado River Compact) were met . . . only in 59-75% of years for the future climate runs" (Christensen et al., 2004).

Hydroclimatic science experts have used different assumptions in their models and are constantly improving them. Collectively, the body of research on prospective future changes in Colorado River flows points to a future in which warmer conditions across the region are likely to contribute to reductions in snowpack, an earlier peak in spring snowmelt, higher rates of evapotranspiration, reduced late spring and summer flows, and reductions in annual runoff and streamflow. Earlier studies suggested substantial decreases in Colorado River annual flow volumes over the next century; more recent studies have generally projected more modest declines, with a few modeling exercises suggesting increases. It is worth reiterating that Colorado River hydroclimate sensitivity studies that consider only the impacts of future temperature increases all forecast decreases in runoff. Forecasts show greater variability when considering possible future changes in precipitation. Modeling results across the region show little consensus regarding changes in future precipitation amounts or seasonality.

Any future decreases in Colorado River streamflow, driven primarily by increasing temperatures, would be especially troubling because the quantity of water allocations under the Law of the River already exceeds the amount of mean annual Colorado River flows. This situation will become even more serious if there are sustained decreases in mean Colorado River flows. Results from these numerous hydroclimatic studies are not unanimous, and all projections of future conditions contain some degree of uncertainty. Nevertheless, the body of climate and hydrologic modeling exercises for the Colorado River basin points to a warmer future with reductions in streamflow and runoff.

This discussion has centered on the mainstem Colorado River, dominated by its two huge reservoirs capable of storing several years of flow. There is, however, an important caveat in this discussion regarding tributary flows: climate change implications for streamflow and reservoir management of the many individual upper basin tributaries upstream from Lakes Powell and Mead may vary considerably from those for the mainstem because of seasonal, topographic, legal, and physical infrastructure constraints particular to each specific sub-basin.

INSTRUMENTAL RECORD OF COLORADO RIVER STREAMFLOW

A streamflow gage monitors a river's flow at a given geographic site; analyzed collectively, a network of streamflow gages can provide an integrated account of weather and climate fluctuations and Earth surface processes over a watershed. The two fundamental hydrologic variables recorded at a streamflow gage are stage (depth) and flow (discharge). Stage measures the height of the water surface relative to some arbitrary datum, whereas flow is the total volume of water that flows past a given point in a specified period of time (e.g., cubic meters per second).

Because river discharge is difficult to measure accurately and continuously, easier and simpler river stage measurements are made instead. These measurements are converted to discharge values through the use of *rating curves*. Rating curves show the relation between stage and discharge, and must be calibrated from available simultaneous measurements of both quantities for each particular gage station. These rating curves must be revised occasionally to reflect changes that affect the hydraulics in the vicinity of the gage. These changes can occur because of changes in river cross sections that result from scour or deposition of sediment, changes in stream gradients, other changes in stream channel morphology and bank structure and roughness (such as from floods), changes in land use across a watershed, or transbasin diversions. This is one reason why rocky locations are preferred for gages; they remain relatively stable. Stream gaging methods and instrumentation have improved greatly over time. Nevertheless, because of the practical challenges in measuring river stage and flow accurately over long time periods, and because of the many physical changes that take place across a watershed and that affect stage-discharge relations, some degree of inaccuracy is often contained in stream gage readings.

The U.S. Geological Survey (USGS) is responsible for the national network of streamflow gages. Over the past century, many USGS stream gages have been relocated and/or the datum have changed at least once; in addition, methods of measuring streamflow (or river stage) have also changed over time (LaRue, 1916; USGS, 1954). To assess the accuracy of gage records, the USGS publishes accuracy information in annual Water Resources Data Reports rating the data records (part or whole) as "excellent" (95 percent of the daily discharges are within 5 percent of the true value), "good" (within 10 percent), "fair" (within 15 percent), and "poor" (Fisk et al., 2004). Some early water supply papers documenting data revisions and gage changes also include accuracy information. Estimating and revising data may improve the completeness of streamflow records but the data may be neither highly accurate nor may it represent true system dynamics. The records indicate that data accuracy may be reasonable except when flows are estimated; this is an important point, given that many Colorado River flow records are based on estimates.

Direct measurements taken at streamflow gages along the Colorado River, in conjunction with similar data obtained from tributary streams across the basin, constitute an important part of the Colorado River hydrologic knowledge base. The two earliest sets of streamflow records used in Colorado River Compact negotiations were from the Green River at Green River, Utah, and at Green River, Wyoming. These records began in 1894 and 1895, respectively (Table 3-1 lists select Colorado River basin gages).

The best-known Colorado River stream gage record is from Lees Ferry, Arizona, where the USGS has been operating a gaging station since May 8, 1921 (Topping et al., 2003). Lees Ferry was selected as a gaging site because it was readily accessible by automobile and was strategically located with respect to Colorado River hydrology. Discharge readings at Lees Ferry measure the combined runoff from the upper part of the Colorado River basin, which includes the upper Colorado, Green, and San Juan rivers (Topping et al., 2003). Located near (~1 mile upstream) the mouth of the Paria River, Lees Ferry was also located several miles downstream from a proposed dam site in Glen Canyon favored by the Southern California Edison Company. As explained in Chapter 2, the 1922 Colorado River Compact designated Lees Ferry as the hydrologic dividing point between the upper and lower basins. The record from Lees Ferry is the most prominent measured record of Colorado River flows (Figure 3-5).

Figure 3-5 shows annual, natural Colorado River flows at Lees Ferry for water years (October through the following September) 1906-2006. Also shown are the long-term average flow value for 1906-2006 (red line) and a 5-year moving average flow value (black line, plotted at the end of each 5-year interval). The mean annual flow value in this instrumental record is roughly 15 million acre-feet (red line). The drought of the late 1990s and early 2000s—which began in the fall of 1999 (water year 2000)—clearly stands out within the past century, as it represents the lowest 5-year running average discharge in the instrumental record.

With respect to Figure 3-5 it is important to distinguish between natural flows (as shown in the data in Figure 3-5) and depleted flows. Depleted flows reflect actual measurements of stream flows and re-

TABLE 3-1 Select Colorado River Gages

Station ID	Station Name	Period
9011000	Colorado River near Grand Lake, CO	1904-1986
9019500	Colorado River near Granby, CO	1908-present
9034500	Colorado River at Hot Sulphur Springs, CO	1904-1995
9058000	Colorado River near Kremmling, CO	1904-present
9070500	Colorado River near Dotsero, CO	1940-present
9072500	Colorado River at Glenwood Springs, CO	1899-1966
9085100	Colorado River below Glenwood Springs, CO	1966-present
9095500	Colorado River near Cameo, CO	1933-present
9106000	Colorado River near Palisade, CO	1902-1933
9153000	Colorado River near Fruita, CO	1911-1923
9163500	Colorado River near Colorado-Utah state line	1951-present
9180500	Colorado River near Cisco, UT	1913-present
9188500	Green River at Warren Bridge, near Daniel, WY	1931-present
9191000	Green River near Daniel, WY	1912-1932
9216500	Green River at Green River, WY	1895-1939
9217000	Green River near Green River, WY	1951-present
9315000	Green River at Green River, UT	1894-present
9335000	Colorado River at Hite, UT	1947-1958
9379500	San Juan River near Bluff, UT	1914-present
9379910	Colorado River below Glen Canyon Dam, AZ	1965-present
9380000	Colorado River at Lees Ferry, AZ	1921-present
9402500	Colorado River near Grand Canyon, AZ	1937-present
9421500	Colorado River below Hoover Dam, AZ-NV	1934-present
9423000	Colorado River below Davis Dam, NV-AZ	1905-1907, 1949-present
9424000	Colorado River near Topock, AZ	1917-1982
9429490	Colorado River above Imperial Dam, CA-AZ	1934-present
9429500	Colorado River below Imperial Dam, CA-AZ	1934-present
9521000	Colorado River at Yuma, AZ	1904-1965, 1983
9521100	Colorado River below Yuma Main Canal WW at Yuma, AZ	1963-present
9522000	Colorado River at NIB AB Morelos Dam near Andrade, CA	1950-present

SOURCE: Harding (2006).

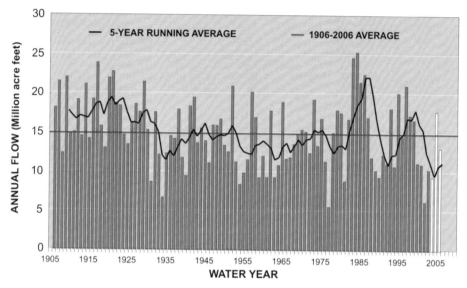

FIGURE 3-5 Natural Colorado River flows at Lees Ferry, AZ, 1906-2006.
NOTE: Black line is 5-year running average and is plotted at the end of 5-year interval. Water years are denoted by the ending year. White bars for 2004-2006 represent preliminary estimates.
SOURCE: Data for 1906-2003 from http://www.usbr.gov/lc/region/g4000/NaturalFlow/index.html.
Values for 2004-2006 are preliminary estimates from J. Prairie, USBR, personal communication, 2006.

flect the actual amount of water flowing past a gage. These flows are typically depleted from their otherwise natural values as a result of upstream diversions (minus return flows), evaporative losses from reservoirs, bank seepage in reservoirs through porous rock, and other upstream depletions. These depletions can be substantial: for example, estimated annual average evaporation from Lake Powell is on the order of 0.5 million acre-feet, while at Lake Mead it is on the order of roughly 0.8 million acre-feet (USBR, 1977; 1981; 1991; 1998; 2002; 2004). Natural flows, by contrast, are estimates of flows that would have occurred without losses from upstream diversions, reservoir evaporation, and the like. Given the extent of human activity in most rivers across the western United States, records of "natural flows" across the West thus almost always represent estimates and not measured flow values.

The Colorado River, of course, has seen numerous upstream depletions and diversions. Water was diverted from the Colorado's headwaters as early as 1892 (Fradkin, 1984). These early depletions resulted in a roughly 10-15 percent reduction in the natural (undepleted) flow of the Colorado River at Lees Ferry up until 1963, when Lake Powell (which is impounded by Glen Canyon Dam) began filling (Ferrari, 1988). From 1963 through 2003, Lees Ferry flows are assumed to be approximately the sum of the flow volumes of the principal rivers—the Colorado, the Green, and the San Juan—that flow into Lake Powell. Thus, the record in Figure 3-5 represents estimated natural flows and it contains uncertainties related to inaccuracies both in measurements and in estimations of natural flows from various depletions.

Several noteworthy hydrologic periods are reflected in the Lees Ferry gage record. The time period used in Colorado River Compact negotiations—1905-1922—included some particularly wet years. This wet period had important ramifications for the Colorado River Compact and its water obligation and allocation agreements. The Compact framers were interested in the river's mean long-term flow. Data records for over two-thirds of the gages used in the negotiations did not begin until 1905 or later; several very low flow years prior to 1905 thus were not fully reflected at that time (Hundley, 1986). Transcripts of Colorado River Compact negotiations describe occasions when Colorado River Commission representatives expressed concern about potentially overly optimistic estimates of annual flow for the Colorado River, perhaps in recognition of some of the low flows prior to 1905 (*http://wwa.colorado.edu/resources/colorado_river/compact/*). A mean annual flow value for the period of record at Yuma, Arizona, 16.4 million acre-feet, was eventually accepted. As now documented in the gaged record, the 1905-1922 period contained the highest long-term annual flow volume in the 20th century, averaging 16.1 million acre-feet per year at Lees Ferry. Other important hydrologic periods reflected in the Lees Ferry record are drought conditions during the Dust Bowl period of the 1930s, drought in the 1950s and in the 1960s, a pronounced regional drought in 1976-1977, and El Niño conditions in 1983-1984.

When the Colorado River Compact was being negotiated, participants had only two to three decades of stream gage data, and only from a small number of stations. Over time, the Lees Ferry gaged record accumulated more and more years of flow data, and the

BOX 3-1
The Colorado River Stream Gaging Network

Over time the instrumental record of Colorado River flows has been augmented with other hydroclimate data and techniques, such as statistically based models for estimating streamflow. Nevertheless, the U.S. Geological Survey (USGS) system of streamflow gages across the Colorado River basin remains a fundamental component of reliable information on flows of the Colorado and its tributary streams. Despite the value of stream gage data—especially from gages that have been measuring streamflow at a given site for many decades—the level of support for these gages has not always been consistent and has seen periods of decline.

The USGS streamflow gaging network shrank from 1980 through the late 1990s because of constraints in funding from both the USGS and from its many partners who also provide resources for this network. In particular, from 1980 to 2000, the USGS stream gaging network lost about 1,790 stream gages that had at least 30 years of record (*http://water.usgs.gov/nsip/ 2007_budget.html*). Because of congressional concern, in 1999 the USGS developed the National Streamflow Information Program (NSIP), a plan to stabilize and modernize the network and provide a defined "backbone" of high-priority stream gages critical to public safety and long-term water resource assessment. The NSIP calls for federal investments in a core network of stream gages that meet national needs and to modernize and improve the reliability of the network. Congress provided significant new funding—approximately $9 million—to begin the implementation of NSIP in Fiscal Year 2001 (*http://water.usgs.gov/nsip/2007_budget.html*). This infusion of funding temporarily reversed the decline of the network and resulted in an additional ~500 stream gages. The loss of long-record gages declined from an average of about 100 per year in the 1990s to less than 30 in 2001. However, in 2004 and 2005 there were more losses of gages (a net loss of about 150 gages), and over 120 long-record stream gages were discontinued in 2004 (*http://water.usgs.gov/nsip/2007_budget.html*).

Although sophisticated techniques are being employed to help augment data gathered from stream flows, the network of gaging stations across the Colorado River basin (especially gages with long-term flow data) provides information that is crucial in describing trends and effects of land use changes, water use changes, and climate changes on the hydrologic system. It thus is important that this gaging network across the basin be maintained and, where possible, expanded.

network of gaging stations also expanded. (That network has not expanded continuously, however, and efforts to add new gaging stations have faced budgetary and other challenges. Box 3-1 discusses the maintenance and value of the USGS streamgaging network.) From the vantage point of the early 21st century, there is now a greater appreciation that the roughly 100 years of flow data within the Lees Ferry gage record represents a relatively small window of time of a system that is known to fluctuate considerably on scales of decades and centuries.

An important question that accompanies the use (exclusively) of the gaged record for river basin planning decisions is how representative the past record is of expected future conditions. To examine the issue of how well the historic, gaged record represents longer-term flow patterns, scientists employ proxy methods. As explained earlier in this report, these proxy data act as stand-ins for instrumental data but cover much longer time spans. As it happens, trees are sensitive to the same climatic elements that cause streamflow to fluctuate, and they live long enough to retain this history within their annual growth rings, in both living and dead trees. The arid climates of the southwestern United States and intermountain Rockies fortunately preserve evidence of past precipitation extraordinarily well. The following section discusses the science of dendrochronology and how this field is used to reconstruct past, long-term Colorado River flows.

TREE-RING SCIENCE AND RECONSTRUCTED STREAMFLOW RECORDS

Records of streamflow measured by gages are limited to little more than the last 100 years. Natural recorders of hydrologic conditions can be used to extend estimates of streamflow back in time to lengthen gaged records and provide a longer context for assessing flow characteristics of the 20th and early 21st centuries. Tree rings are the best source of high-resolution, precisely dated proxy records of hydroclimatology over the past several centuries and they have proven useful for reconstructing a range of hydroclimatic variables, including temperature, precipitation, and streamflow (Woodhouse and Meko, 2007). Although the record of past hydroclimatic variability may not be replicated in the future, the extended records are useful for documenting a broader range of natural variability than provided by the gaged record alone. This section reviews basic concepts underlying tree-ring-based streamflow reconstructions and the uncertainties inherent in them. It includes a comparison of reconstructions of upper Colorado River basin streamflow and discusses the features of the most recent Lees Ferry reconstructions, along with implications for sub-basin flow relationships.

Scientific Basis of Streamflow Reconstructions

Tree-ring reconstructions of past hydrologic conditions are based on the principles of dendrochronology, the science and study of dated tree rings (Fritts, 1976). Dendrochronology allows the dating of tree rings to the exact year of formation by matching ring-width patterns from tree to tree using a technique known as cross dating. This precise dating is critical because annual increments of tree growth are directly calibrated with annual measurement of hydroclimatic variability in the streamflow reconstruction process. Cross dating is possible because trees that are limited in growth primarily by climate will share a similar pattern of ring-width variations with other trees across a climatically homogeneous region.

In the Colorado River basin, coniferous tree species growing at lower elevation and, in particular, stands of trees on well-drained slopes with southern exposures have been shown to be well suited for reconstructions of annual streamflow (Hidalgo et al., 2000; Meko and Graybill, 1995; Michaelsen et al., 1990; Schulman, 1956; Smith and Stockton, 1981; Stockton, 1975; Stockton and Jacoby, 1976; Woodhouse and Lukas, 2006; Woodhouse et al., 2006). These coniferous species include ponderosa pine (*Pinus ponderosa*), pinyon pine (*Pinus edulis*), and Douglas fir (*Pseudotsuga menziesii*). A typical life span for trees within these three species is 300-500 years, and some individuals can live to be over 800 years old. Annual tree growth at these moisture-stressed sites appears to depend on soil moisture in the early part of the growing season (Meko et al., 1995). Climatic conditions that affect spring and early summer soil moisture include antecedent moisture conditions in the prior late summer and fall, and winter snowpack. This set of conditions is also important for surface water flows. Annual (water year) streamflow thus is often highly correlated with the annual tree growth of these moisture-sensitive species (see Meko et al. [1995] and Meko [2005] for more detailed discussions on methods for assessing relationships between annual tree-ring growth and streamflow).

To generate streamflow reconstructions, trees are sampled with an increment borer at collection sites based on the factors described above that affect tree ring growth. Sample replication at individual sites is important, and two cores are collected from each of 15-40 different trees per site. Cores from each site are cross-dated, measured, standardized (e.g., the size/age trend is removed), and combined into tree-ring site chronologies (Cook and Kairiukstis, 1990; Stokes and Smiley, 1968), which are the basis of a streamflow reconstruction. Tree-ring chronologies are calibrated with gage data to develop a reconstruction model. Several statistical approaches, typically based on multiple linear regression, have been employed to develop these models (Loaciga et al. [1993] provide a review of these approaches). Reconstruction models are evaluated with a suite of statistics that quantifies the variance explained in the gaged record by the reconstruction, and the uncertainty related to the unexplained variance.

Reconstructions are validated by testing the model on data not used in the calibration, to ensure that the model is not tuned specifically to the calibration data, but performs well on independent data as well (Cook and Kairiukstis, 1990; Fritts, 1976; Loaciga et al., 1993). The model is applied to the full length of the chronologies to generate an extended record of flow. In applying these models back in time, the assumption is made that the relationship between tree growth and climate in the calibration period also existed in the past, while recognizing that conditions of the past were not necessarily the same as in the instrumental period (Fritts, 1976).

Uncertainties in Streamflow Reconstructions

Considering that reconstructions are only estimates of flow, uncertainties in these reconstructions derive from several different sources. The fact that trees are imperfect recorders of hydrologic variability is an inherent source of uncertainty and is reflected in the inability of tree ring-based models to account for 100 percent of the variance in the gaged record (e.g., Brockway and Bradley, 1995). This also makes a direct comparison between reconstructed and gaged values inappropriate unless this uncertainty is considered. The precision with which tree-growth rings can be used to estimate past flows is quantified by the statistical model generated in the calibration, and a measure of the error in the reconstruction model can be used to describe model confidence intervals. However, this is only one source of uncertainty. Other sources include changes in tree-ring sample numbers over time, which affects the strength of the common (hydroclimatic) signal in the reconstruction (Cook and Kairuikstis, 1990; Wigley et al., 1984). Uncertainties can also derive from the preservation of low-frequency (multidecadal to centennial scale) information in the tree-ring data, which is limited by the lengths of the individual tree-ring series and how these series were standardized to remove the biological growth curve (Cook et al., 1995). There is also some degree of uncertainty because of the quality of the gage record used for the calibration and how that may vary over time. In addition, most reconstructions better replicate dry extremes than wet extremes (Michaelsen, 1987). Reconstructed flows that are higher or lower than the range of values in the gage record often reflect tree-ring variations beyond the range of variations in the calibration period, and may be less reliable than indicated by regression results (Graumlich and Brubaker, 1986; Meko and Graybill, 1995; Meko et al., 1995).

Dendrochronologists have long acknowledged and reported the model error in reconstructions, although error bars have not typically been presented with reconstructions, which would reinforce the probabilistic nature of the reconstruction values. A variety of techniques are used, with some currently under development, to identify and quantify other sources of uncertainty (Meko et al., 2001; Woodhouse and Meko, 2007). An approach to systematically quantify the amount of error attributed to each of these sources, however, has not yet been developed.

Reconstructions of Colorado River Flows at Lees Ferry, Arizona

As methods for tree-ring-based reconstructions have evolved, the set of streamflow data from the Lees Ferry gage has been a focus of reconstruction studies. Several reconstructions for Lees Ferry flow have been generated, first by Stockton and Jacoby (1976), followed by Michaelsen et al. (1990), Hidalgo et al. (2000), and Woodhouse et al. (2006). Stockton and Jacoby (1976), Michaelsen et al. (1990), and

Hidalgo et al. (2000) used similar networks of tree-ring data, with at least 30 percent of the chronologies shared and with a common end date of 1963. Woodhouse et al. (2006) used a new network of tree-ring data, ending in 1995. All four studies used different gage data for calibration, and Stockton and Jacoby (1976) used two different sources of gage data, illustrating the difference the gage records can make in the final reconstruction. The number of years for calibration also varied from 49 to 90 years. The reconstructions also included some differences in the statistical treatment of the tree-ring data and statistical approaches to the calibration (see Table 3-2).

The resulting reconstructions differ in some respects. Given that these studies employed different data sets, assumptions, and methods, some differences across results are to be expected. All these reconstructions, however, share similar key features with respect to the timing and duration of major wet and dry periods. These reconstructions, as depicted in Table 3-2 and shown in Figure 3-6, support the following points:

1. Long-term Colorado River mean flows calculated over these periods of hundreds of years are significantly lower than both the mean of the Lees Ferry gage record upon which the Colorado River Compact was based and the full 20th century gage record (Woodhouse et al., 2006).

2. High flow conditions in the early decades of the 20th century were one of the wettest in the entire reconstruction.

3. The longer reconstructed record provides a richer basis from which to assess the range of drought characteristics that have been experienced in the past, revealing that considerably longer droughts have occurred prior to the 20th century.

These three points have important implications for water management decisions for the Colorado River basin and are revisited in the Commentary section at the end of this chapter.

TABLE 3-2 Lees Ferry Reconstructions

Reconstruction	Calibration Years	Source of Gage Data	Chronology Type[c]	Regression Approach[d]	Variance Explained	Reconstruction Years	Long-Term Mean[e] (MAF)
Stockton and Jacoby (1976)	a. 1899-1961 b. 1914-1961 c. 1914-1961 Average of a and b	Hely (1969) Hely (1969) UCRSFIG (1971)	Standard Standard Standard	PCA with lagged predictors	0.75 0.78 0.87	1512-1961 1512-1961 1511-1961 1520-1961	14.15 13.9 13.0 13.4
Michaelsen et al. (1990)	1906-1962	Simulated flows[a]	Residual	Best subsets	0.83	1568-1962	13.8
Hidalgo et al. (2000)	1914-1962	USBR, see Hidalgo et al. (2000)	Standard	Alt. PCA with lagged predictors	0.82	1493-1962	13.0
Woodhouse et al. (2006)		USBR[b]					
Lees-A	1906-1995		Residual	Stepwise	0.81	1490-1997	14.7
Lees-B	1906-1995		Standard	Stepwise	0.84	1490-1998	14.5
Lees-C	1906-1995		Residual	PCA	0.72	1490-1997	14.6
Lees-D	1906-1995		Standard	PCA	0.77	1490-1998	14.1

[a] Simulated flows developed from the U.S. Bureau of Reclamation (USBR) Colorado River Simulation System.
[b] J. Prairie, USBR, personal communication, 2004.
[c] Standard chronologies contain low-order autocorrelation related to biological persistence; residual chronologies contain no low order autocorrelation.
[d] Regression approach: PCA is principal components analysis (regression). Best subsets is multiple linear regression, using Mallow's Cp to select best subset. Alternative PCA used an algorithm find the best subset of predictors on which to perform PCA for regression. Stepwise is forward stepwise regression.
[e] Long-term mean based on 1568-1961 except for Michaelsen et al. (1990), which is based on 1568-1962.

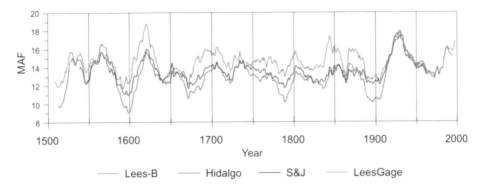

FIGURE 3-6 Colorado River annual streamflow reconstructions, Lees Ferry, AZ (smoothed with a 20-year running mean).
NOTE: Year plotted is the last year in the 20-year mean.
SOURCE: Lees-B (standard chronologies, stepwise regression) from Woodhouse et al. (2006); Hidalgo et al. (2000); S&J from Stockton and Jacoby (1976; average of two models); Lees Gage is gage record, 1906-1995, J. Prairie, USBR, personal communication, 2004.

Differences among the Reconstructions

The most obvious difference among the reconstructions is the long-term mean, a measure with implications for long-term water allocation decisions. The reconstructions based on the calibration periods that end in 1961 or 1962 generally have lower long-term means than more recent reconstructions with a calibration period that ends in 1995 (Table 3-2). A second noticeable difference is the magnitude of the high and low flow periods, which vary between all reconstructions to some degree.

Some differences in the Lees Ferry reconstructions may be attributed to the tree-ring and gage data, including the length of the calibration period. Differences may also result from choices made in statistical methods when processing tree-ring data, which can affect the characteristics of the chronology and, in turn, affect the reconstruction (see Meko et al. [1995] and Woodhouse and Meko [2007] for details on the treatment of tree-ring data). In the Lees Ferry reconstructions, Stockton and Jacoby (1976) and Hidalgo et al. (2000) used chronologies that retained the biological persistence (standard chronologies), which is the tendency for a tree's growth in one year to be associated with growth in a following year due to biological processes. In contrast, Michaelsen et al. (1990) used chronologies in which this biological persistence was statistically removed. Woodhouse et al. (2006) tested models using both types of chronologies. Different results may also arise from the statistical approach used in the calibration process and can stem from the inclusion of "lagged predictors" (tree-ring chronologies lagged forward and backward several years relative to the gage record) and details of regression methods used (see Woodhouse et al. [2006] for more information on statistical methods used in dendrochronology research).

The sensitivity of the resulting reconstructions to some of these statistical treatments and approaches has been tested for Lees Ferry (Woodhouse et al., 2006). Results from this study indicate that different types of chronologies (standard versus residual) can have an influence on the skill of the reconstruction in replicating some of the time-series characteristics of the gage record, and persistence of low flows may be heightened with standard chronologies (Woodhouse et al., 2006). The use of different modeling approaches in model calibration was not an obvious source of differences. In addition, the length of the calibration period did not appear to be critical, as calibrating a model on a shorter time period (1914-1961 versus 1906-1995) resulted in a similar reconstruction (Woodhouse et al., 2006).

In summary, differences among Lees Ferry reconstructions can likely be attributed to several factors. There are some indications that periods of persistent low flows may be accentuated using standard chronologies and/or lagged predictors, but the sources of the differences in long-term mean are not yet clear. Additional studies will be needed to more accurately assess the impact of the different sets of chronologies and gage records on the final reconstructions. As to which reconstruction might be the most accurate or "best," reconstructions with the longest calibration period are statistically more robust (i.e., exhibiting similar results when tested with different models), particularly considering that the recently recalibrated gage record from 1906-1995 is assumed to be the most accurately estimated natural flow data. Within the set of reconstructions calibrated on the longest period, however, there is no clearly superior solution, with each reconstruction containing strengths and weaknesses (e.g., match in persistence in the gage record, over/underestimation of decadal-scale low flows; Woodhouse et al., 2006).

Analyzing Reconstructed Colorado River Flow

The extended record of streamflow provided by the tree-ring reconstructions is useful for assessing the characteristics of the gage record in a long-term context and for examining low-frequency (multidecadal-scale) behavior of flow, which is not possible with the shorter gage record. Questions relevant to drought and management in the upper Colorado River basin that can be addressed are:

1. How does the drought of the early 2000s compare to other past droughts of similar duration?
2. Have longer periods of drought occurred? and
3. What is the character of decadal-scale variability over time compared to the 20th century?

When early 2000 drought conditions are assessed as a 5-year (2000-2004) mean value, the reconstruction indicates one period—1844-1848—with a lower mean value, but several additional periods with a fairly high probability of being lower as well (Woodhouse et al., 2006). The Lees Ferry gage record contains no periods of below median flow that lasted more than 5 consecutive years. In the Lees Ferry tree-ring-based reconstruction, however, longer periods of below-median flow have occurred, including periods of up to 10 and 11 years. The reconstruction also reflects the nonstationarity—or changes in the values of decadal-scale means—of flow over decadal time scales (Figure 3-6).

Colorado River Sub-Basin Relationships and Circulation

In addition to the record of upper Colorado River flow at Lees Ferry, reconstructions can provide information about long-term hydroclimatic variability within Colorado River sub-basins. Along with Lees Ferry, flow records at gages on major tributaries of the upper Colorado River—the Green River, the San Juan River, and Colorado River mainstem (i.e., before it joins the Green and San Juan rivers, which was historically known as the Grand River)—have been reconstructed (Woodhouse et al., 2006). A comparison of reconstructions for these tributaries suggests that major multiyear droughts and multidecadal dry periods impact the entire basin, although the relative magnitude may vary spatially. Similarly, research that examined reconstructions of several tributaries of the lower Colorado River basin in Arizona—in the Salt and Verde River basins—found droughts (and wet events) in the upper Colorado and Salt-Verde River basins to be concurrent more often than not (Hirschboeck and Meko, 2005). Details of the primary mechanisms that influence upper Colorado River basin climate and hydrology at multidecadal time scales are not yet clear. Studies of extended periods of streamflow, however, considered along with other high-resolution climate reconstructions, have the potential to increase scientific understanding of the links between ocean/atmosphere circulation and Colorado River basin water supply.

COMMENTARY

A steady warming trend of about 2°F has been under way over the past three decades across the Colorado River basin. Results from several different climate modeling experiments indicate that future temperatures will continue to rise across the Colorado River basin. Projections of annual precipitation changes from these same models exhibit a range of results, most of them approximately centering around present values. The models project a tendency for increases in winter precipitation of about the same magnitude as decreases in summer precipitation. Higher temperatures will cause higher evaporative losses from snowpack, surface reservoirs, irrigated land, and land cover surfaces across the river basin. Hydrologic modeling studies of future Colorado River runoff exhibit a variety of results, and such forecasts always reflect some degree of uncertainty. Collectively, however, these studies indicate that future Colorado River streamflow will decrease with increasing future temperatures.

The 20th century saw a trend of increasing mean temperatures across the Colorado River basin that has continued into the early 21st century. There is no evidence that this warming trend will dissipate in the coming decades; many different climate model projections point to a warmer future for the Colorado River region.

Modeling results show less consensus regarding future trends in precipitation. Several hydroclimatic studies project that significant decreases in runoff and streamflow will accompany increasing temperatures. Other studies, however, suggest increas-

ing future flows, highlighting the uncertainty attached to future runoff and streamflow projections. Based on analysis of many recent climate model simulations, the preponderance of scientific evidence suggests that warmer future temperatures will reduce future Colorado River streamflow and water supplies. Reduced streamflow would also contribute to increasing severity, frequency, and duration of future droughts.

In the context of multidecadal and multicentury hydroclimatic patterns across the Colorado River region, the Lees Ferry gaged record represents a chronologically limited sliver of information. Paleoclimate-based reconstructions of Colorado River streamflow have become of great interest to water managers across the region because, instead of 100 years of Colorado River flows, the reconstructions provide estimates of hundreds of years of flows. The first tree-ring-based reconstruction was developed in the 1970s and has been followed by several other studies using similar tree-ring data. Although the various reconstructions are not perfectly congruent, this is not unexpected given that the reconstructions were independently developed by scientists using different data sets and relying upon differing assumptions and statistical methods. Nonetheless, the reconstructions exhibit broad agreement in several important respects: they replicate similar past wet and dry periods; they suggest that the Colorado River's long-term mean annual flow is less—ranging from 13 to 14.7 million acre-feet—than 15 million acre-feet (the mean annual value based on the Lees Ferry gaged record); they show that the 1905-1920 period was one of the wettest periods in the past several centuries; and, they indicate multiple drought periods that were more persistent and severe than droughts reflected in the gaged record. Past climates may not necessarily be replicated in the future but reconstructions of past flows provide information that, when used in concert with projections of future climate, can offer valuable guidance to aid future water resources planning.

Although much remains to be learned about the drivers of hydroclimatic variability in the basin—particularly those that operate at multidecadal and longer time scales—the scientific foundation underlying contemporary understanding of Colorado River streamflow patterns has evolved markedly during the past 50 years. Whereas in the mid-1950s that foundation relied almost solely upon gaged flow records, today it consists of a more sophisticated understanding and modeling of the global climate system, better temperature data from the Colorado River region and across the world, paleoclimate studies and streamflow reconstructions, and a longer record of gaged river flows. Assessed collectively, this body of knowledge invalidates any assumption that Colorado River flows vary around an annual mean value that is static and unchanging.

For many years, scientific understanding of Colorado River flows was based primarily on gaged streamflow records that covered several decades. Recent studies based on tree-ring data, covering hundreds of years, have transformed the paradigm governing understanding of the river's long-term behavior and mean flows. These studies affirm year-to-year variations in the gaged records. They also demonstrate that the river's mean annual flow—over multidecadal and centennial time scales, as shown in multiple and independent reconstructions of Colorado River flows—is itself subject to fluctuations. Given both natural and human-induced climate changes, fluctuations in Colorado River mean flows over long-range time scales are likely to continue into the future. The paleoclimate record reveals several past periods in which Colorado River flows were considerably lower than flows reflected in the Lees Ferry gaged record, and that were considerably lower than flows assumed in the 1922 Colorado River Compact allocations.

Multicentury, tree-ring-based reconstructions of Colorado River flow indicate that extended drought episodes are a recurrent and integral feature of the basin's climate. Moreover, the range of natural variability present in the streamflow reconstructions reveals greater hydrologic variability than that reflected in the gaged record, particularly with regard to drought. These reconstructions, along with temperature trends and projections for the region, suggest that future droughts will recur and that they may exceed the severity of droughts of historical experience, such as the drought of the late 1990s and early 2000s.

Data from the gaging station at Lees Ferry, Arizona, represent the best-known Colorado River measured flow record. As flow data accumulated over time at Lees Ferry, it became clear that 1905-1920— the period upon which Colorado River allocations were ascribed— was significantly wetter than average. It has also become evident that the river's average annual flow is considerably less than the approximately 16.4 million acre-feet figure used by Colorado River Compact

Climate and Hydrology of the Colorado River Basin Region

negotiators. For many years the 20th century gage record of Colorado River flows represented the best understanding of the river's year-to-year hydrologic variability. Despite the value of data from these gage records—especially from sites that have accumulated data for several decades—support for the USGS system of stream gages has not always been steady and has seen some past periods of decline. Today, science-based knowledge of the river's flows and the basin's climate systems has become more sophisticated. Nevertheless, the gage record of river flows will remain an important source of information for scientists and water resources planners.

Measured values of streamflow of the Colorado River and its tributaries provide essential information for sound water management decisions. Loss of continuity in this gaged record would greatly diminish the overall value of the existing flow data set, and once such data are lost they cannot be regained. The executive and legislative branches of the U.S. federal government should cooperate to ensure that resources are available for the USGS to maintain the Colorado River basin gaging system and, where possible, expand it.

4

Prospects for Conserving and Extending Water Supplies

The history of western U.S. water development has been one in which storage reservoirs and related conveyance facilities were constructed to provide water supplies to cope with occasional drought, as well as to encourage population growth and economic development. This strategy has been complemented by a variety of means for increasing supplies and better managing demands: groundwater supplies have been tapped, irrigation practices have been refined and improved, some states and cities have adjusted landscaping practices, and there have been efforts at weather modification. More recently, both technical and legal aspects of groundwater storage methods have become more sophisticated and increasingly applied. As described in Chapter 2, the strategy of building additional surface water storage capacity is encountering physical, economic, and political limits. As more traditional water projects have become less viable, and as water demands continue to grow, federal, state, and municipal water managers across the West are considering a new water project prototype that entails nonstructural measures such water conservation, water use technologies, xerophytic landscaping, groundwater storage, and changes in water pricing policies. This chapter reviews a variety of techniques and initiatives that have been and are being explored as means to augment and extend water supplies across the Colorado River basin.

LARGE-SCALE RESERVOIRS AND INTER-BASIN TRANSFERS

Between the 1930s and the 1970s, many multipurpose dams and reservoirs were constructed in the Colorado River basin in an effort to smooth natural variations in the river's flows and to store flood waters for use during drier periods. The prototype of these structures was Hoover Dam. The Colorado River Storage Project (CRSP) of 1956 represented another water development milestone as it authorized construction of Glen Canyon Dam (in Arizona near the Utah border), Flaming Gorge Dam (on the Green River in Utah near the Wyoming border), Navajo Dam (on the San Juan River in New Mexico near the Colorado border) and the multidam Wayne N. Aspinall Storage Unit (on the Gunnison River in western Colorado; see *http://www.usbr.gov/dataweb/html/crsp.html*). The CRSP represented the zenith of large-scale dam construction across the basin. Following the 1956 passage of CRSP and the construction of its authorized projects, new factors in the planning of western water resources began reducing the prospects for new projects. A burgeoning environmental movement in the post-World War II era raised awareness of environmental changes wrought by dams, leading in part to the defeat of proposals to build dams at Echo Park in Dinosaur National Monument (in the 1950s) and at Bridge and Marble Canyon near Grand Canyon National Park in the 1960s (Nash, 1967; Reisner, 1986). The trend toward fewer traditional, structural western water projects continues today, as the best sites for storage reservoirs have been developed and as concerns have grown over environmental impacts of large dams, both in the Colorado basin and elsewhere (see WCD, 2000). Some water storage and delivery projects were completed in the 1980s and 1990s, perhaps most notably the Central Arizona Project in 1992, but the declining trend of the viability of traditional water projects has been clear.

In addition to environmental and other concerns related to large dams, traditional water projects today face a more stringent series of planning and feasibility studies and other obligations than in the past, which can entail literally decades of project planning and related activities. (Box 4-1 discusses the Animas-La Plata project in southwestern Colorado, which is an example of the complexities that surround contemporary dam authorization, appropriation, and construction.) In efforts to augment water supplies, some basin states and

BOX 4-1
The Animas-La Plata Project

Congress authorized the Animas-La Plata project in 1968, calling for a multipurpose dam project to serve a range of agricultural, municipal, and industrial uses in southwestern Colorado. Today, 37 years after project authorization, the Bureau of Reclamation's Animas-La Plata project is under construction. Although scheduled for construction in the early 1980s, discussions were initiated to achieve a negotiated settlement of water rights claims of the Southern Ute Indian and Mountain Ute tribes in southwestern Colorado. Following negotiations, a settlement of water rights claims held by these tribes was agreed to in a Final Settlement Agreement, signed on December 10, 1986.

In 1990, the U.S. Fish and Wildlife Service issued a draft biological opinion regarding the federally endangered Colorado pike minnow and how it might be affected by Animas-La Plata. A final biological opinion was issued in 1991, which allowed for construction of several Animas-La Plata project features, but limited annual project depletions to 57,100 acre-feet while an endangered fish recovery program was conducted. After the U.S. Bureau of Reclamation was authorized to initiate construction, several challenges were made to the completeness of Reclamation's 1980 final environmental impact statement, and in 1992 legal actions brought by environmental organizations halted construction. Reclamation worked with the Fish and Wildlife Service to address new biological information, and in 1996 the Service issued a biological opinion with a reasonable and prudent alternative limiting project construction to features that would initially result in an average annual water depletion of 57,100 acre-feet. Construction of the Ridges Basin Dam, the centerpiece of the Animas-La Plata Project that will impound 120,000 acre-feet, began in 2005. The reservoir, to be named for former Colorado Senator Ben Nighthorse Campbell, is expected to be filled in 2011.

The history of the Animas La-Plata project reflects how difficult it can be for western water projects to move from planning to construction. The process today is far more complicated than during the 1950s and 1960s. Although future storage dams may be built within the Colorado River basin, the Animas-La Plata experience offers little evidence that they will be built quickly.

SOURCES: *http://www.usbr.gov/uc/progact/animas/background.html*; Rodebaugh (2005).

Prospects for Conserving and Extending Water Supplies

municipalities may still wish to pursue the option of constructing a new water storage reservoir(s). Viable prospects for new project construction in the near to medium term, however, are limited: "Except for the Central Utah Project, as recently modified by Congress, and perhaps the Animas-La Plata Project, it seems unlikely that other major water storage facilities will be constructed in the Colorado River Basin in the foreseeable future" (MacDonnell et al., 1995). Although a diversion dam on the Virgin River has been discussed, there is no current proposal to build such a project, and it is one of the few dam projects that has even been discussed in the basin in recent years.

An interesting chapter in the history of efforts to augment Colorado River basin water supply storage involves various plans to import water from outside the basin. The most ambitious of these was the North American Water and Power Alliance (NAWAPA), an engineering scheme proposed in 1964 by the Ralph M. Parsons Company of Pasadena, California. The plan envisioned moving large quantities of water from water-rich regions of Alaska and the Canadian Yukon to the arid western United States through a complex system of reservoirs, tunnels, pumping stations, and canals. Dams were also to generate hydropower, sales of which were to help finance project construction. The Parsons Company 1964 cost estimate was $80 billion, adjusted to $130 billion in 1979. The price tag in today's dollars would undoubtedly be in the hundreds of billions. Political and environmental objections would also impede, and likely block, attempts to revive even a scaled-down version of the NAWAPA scheme. Similarly, prospects of towing icebergs south from Alaska or other arctic regions to augment Colorado River water supplies are equally unrealistic. Declining prospects for traditional water supply projects are perhaps more correctly seen not as an end to "water projects" but as part of a shift toward nontraditional means for enhancing water supplies and better managing water demands. The following sections of this chapter examine some nonstructural and nontraditional means of augmenting water supplies.

CLOUD SEEDING

Weather modification, including cloud seeding to increase rainfall and suppress hail, has long generated interest among scientists, public officials, and private practitioners in a dozen or more nations. Cloud seeding has been studied and practiced in the United States for at least five decades. Over this period, research investment by agencies of the federal government has waxed and waned. Early experiments conducted by the U.S. Weather Bureau in the late 1940s showed sufficient promise that federally sponsored efforts were scaled up in the 1950s with programs overseen by the Weather Bureau, the U.S. Air Force, and the National Science Foundation, all of which supported cloud seeding research into the 1960s and 1970s. The mid-1970s marked a high point of federal support for cloud seeding, and the National Weather Modification Act of 1976 spurred federal research efforts and mandated a Department of Commerce Weather Modification Advisory Committee to coordinate research among federal agencies. In this same time frame, assessments were made of scientific progress made over the preceding decade and a half. The assessments include a series of reports from both the National Research Council (NRC) and the National Science Board that concluded that experimental evidence for cloud seeding had not yet definitively established its scientific efficacy (NRC, 1964, 1966, 1973; NSB, 1966). The National Research Council subsequently (in 2003) issued a report on the prospects of cloud seeding and other weather modification techniques, concluding that:

> There is still no convincing scientific proof of the efficacy of intentional weather modification efforts. In some instances there are strong indications of induced changes, but this evidence has not been subjected to tests of significance and reproducibility. This does not challenge the scientific basis of weather modification concepts. Rather, it is the absence of adequate understanding of critical atmospheric processes that, in turn, lead to a failure in producing predictable, detectable and verifiable results (NRC, 2003).

In 2004 the Weather Modification Association (WMA) assessed the NRC report from the perspective of those involved in operational weather modification (Orville et al., 2004). This review supported many of the NRC report's recommendations but also included some criticisms; specifically, the WMA claimed that the NRC report did not adequately account for recent field applications for precipitation enhancement and hail suppression. Since the NRC and WMA reports were issued, some scientists have sought common ground with operators to develop a cloud seeding program that would include scientifically controlled watershed experiments (Garstang et al., 2004).

Federal support for cloud seeding research has generally declined since the mid-1970s. Nevertheless, several parties and states in the Colorado River basin maintain a strong interest in the prospects of cloud seeding to increase precipitation. For example, in a 2005 letter to the Secretary of the Interior, the Governor's Representatives on Colorado River Operations sought to work with the Department of the Interior "to implement a precipitation management (cloud seeding) program in the basin (both Upper and Lower)" (Governors, 2005). In light of the stress on federal funding for discretionary expenditures, a renewed large-scale, federally led weather modification initiative does not appear likely (AAAS, 2006). For the foreseeable future, weather modification experiments and operations will depend mainly on funding from state governments, local communities, and private-sector entities (e.g., utility companies).

Six of the seven Colorado River basin states presently support some type of precipitation or snowpack augmentation operations (WMA, 2005). The most prominent cloud seeding project in the basin may be one sponsored by the Wyoming Water Development Commission. This 5-year project is designed to demonstrate if rainfall and snowpack in the state's mountainous regions can be enhanced (see *http://www.rap.ucar.edu/projects/wyoming/*). Cloud seeding operations are planned in the Wind River Mountains and the Medicine Bow Range/Sierra Madre Mountains. The program is important because of its potential scientific and operational evaluation for the Colorado River basin states and because the 5-year program is to utilize a solid scientific base for the experiments. If the Wyoming pilot trials increase snowpack by 10 percent, the additional yield would, on average, be on the order of 130,000 to 260,000 acre-feet of additional runoff each spring (WWDC, 2006), which would represent a notable increase in water supplies. In addition to the Colorado River basin states, entities such as municipalities and the ski industry are interested in the prospects of augmenting water supplies and snowpacks by cloud seeding. Denver Water, for example, commenced cloud seeding again in 2002 after 20 years of putting its program on hold. Denver Water's cloud seeding program was reinitiated as a response to the 2002 drought and was conducted through March 2003 (see *http://www.denverwater.org/cloud_seeding.html*).

In evaluating the success or benefits of cloud seeding operations, the experience of six decades of experiments and applications that failed to produce clear evidence that cloud seeding can reliably enhance water supplies on a large scale should be kept in mind. Of course, clear evidence is difficult to produce in cloud seeding experiments, as they are not amenable to case-control studies. Furthermore, such experiments are seen by many as being relatively inexpensive even if they do not definitively result in greater precipitation. Given increasing demands for water across the Colorado River basin, cloud seeding is likely to continue to be pursued as a means for augmenting water supply.

DESALINATION

Scientists and engineers, governments, and advocacy groups have long investigated desalination as a means of augmenting freshwater supplies. Most attention has been directed to converting seawater to potable freshwater, while less attention has focused on subterranean and surface brackish water desalination. There have been steady scientific and engineering advances in the technologies of salt water conversion, and several desalination facilities have been constructed. Advances in technology have led to cost reductions, improved efficiency, and an increase in the numbers of desalination plants worldwide. One recent estimate places the total number of desalting plants at 7,500, capable in total of producing several billion gallons of potable water per day (http://www.waterdesalination.com). Nearly half the world's desalinated water production today is in the Middle East; about 15 percent of the world's desalinated water is produced in North America (Wangnick, 2002).

In California there are currently 16 coastal operating or planned desalination facilities (http://www.coastal.ca.gov/desalrpt/dsynops.htm). The San Diego County Water Authority is committed to desalination, and by 2020 expects 15 percent of its supply to come from desalination (http://www.sdcwa.org/manage/sources-desalination.phtml). In addition to interests of municipalities and utilities for coastal desalination facilities, energy companies are operating small desalination plants on offshore oil and gas exploration and production rigs; there are nine rigs with desalination facilities off the coast of California (California Coastal Commission and State Lands Commission, 1999). Not all desalination initiatives have proven fully successful, however. For example, in 1999 water authorities jointly sponsored a privately financed desalination plant at Tampa Bay, Florida, to supplement freshwater supplies for their 1.8 million customers. As of May 2006, the plant was not in operation, being plagued by management and technical problems (Cooley et al., 2006). The experience of the City of Santa Barbara, California represents another prominent example of the challenges associated with large-scale desalination (see Box 4-2).

Recent improvements in desalination technology have led to energy cost reductions per unit of water produced. There is, for example, a variety of membrane technologies such as reverse osmosis, nanofiltration, and ultrafiltration. These all remove salts, dissolved organics, bacteria, and other seawater constituents from salt water (Pankratz and Tonner, 2003). There is also a range of thermal technologies that boil or freeze water, then capture the purified water while the contaminants remain behind.

Energy requirements and costs are important considerations in desalination projects and greatly affect construction plans and decisions in the United States (especially as compared to areas such as the Middle East, where oil and natural gas costs are heavily subsidized). Energy costs notwithstanding, relative production costs have fallen since the early 1990s and the capacity of facilities has risen (AMTA, 2005). In the United States there is some interest in coupling future desalination plants with new power plant production for cogeneration to reduce energy cost in desalination; rising energy costs, however, make it unclear if this trend will continue (Cooley et al., 2006). On the other hand, technical advances may continue and increase desalination efficiency even if energy costs rise. For example, a team led by scientists from the Lawrence Livermore National Laboratory estimates that a membrane system using carbon nanotube-based membranes may be able to reduce future costs of desalination by 75 percent compared to current reverse osmosis membrane technology (Holt et al., 2006).

Longstanding federal research and development programs for desalination have been advanced by a series of congressional authorizations, such as the Water Purification and Desalination Act of 1996 (P.L. 104-298). An NRC report reviewed the Bureau of Reclamation's desalination and water purification program and offered recommendations for program improvement (NRC, 2004). State governments and municipal water districts are also investing in desalination research, development, and demonstration facilities. The Bureau

> **BOX 4-2**
> **Desalination in Santa Barbara**
>
> The City of Santa Barbara, California, relies heavily on rainfall and local groundwater to meet its water supply needs. These sources were impacted by severe drought conditions between 1987 and 1992, which caused sharp declines in local reservoir levels. The water shortage led city officials to consider a new source(s) of water supply, and Santa Barbara residents approved construction of a desalination plant to augment the city's water supplies (they also approved a piped connection to California's State Water Project).
>
> Construction of a reverse osmosis facility began in 1991 and was completed in 1992. The plant successfully produced water during its testing phases, but soon after plant completion, drought conditions in the region subsided. The plant was placed on active standby mode because of the high costs of producing water during nondrought periods. At the same time, the higher costs of water driven by the desalination plant and the connection to the State Water Project contributed to declining water demands. Conservation measures enacted during the 1987-1992 drought, such as low-flow toilets and xerophytic landscaping, contributed to water savings, and per capita demands never rebounded to predrought levels.
>
> The desalination plant today is decommissioned, with a large portion of the plant's infrastructure having been sold to a company in Saudi Arabia. Today, the plant serves as an "insurance policy, allowing the City to use its other supplies more fully" (*http://www.santabarbaraca. gov/Government/Departments/PW/SupplySources.htm?js=false*).
>
> Although the plant is not currently operational, its future will bear watching as California's population continues to grow, as the City of Santa Barbara continues to strive for urban water efficiencies, and as the economics of energy and water production continue to shift.
>
> SOURCE: Cooley et al. (2006).

of Reclamation has recently focused its desalination research and development strategies in three areas: grants to university scientists, studying the feasibility of reopening the Yuma desalination plant, and constructing a test facility at Alamogordo, New Mexico to explore the feasibility of desalting brackish groundwater.

Beyond energy costs, desalination entails several environmental implications. A key barrier to economically viable desalination is disposal of the briny water that is a byproduct of the process. This is especially a problem in areas that do not have access to the ocean, but it can also be problematic for coastal locales. For example, native species in bays and estuaries are impacted by large seawater intake and by discharge of briny concentrates that are byproducts of desalination processes. Drawdown of brackish water in subterranean reservoirs can lead to ground subsidence and/or a lowering of the water table. Regulations and technologies to mitigate adverse possible environmental effects associated with desalination have been and will continue to be implemented by municipalities, states, and the federal government.

Technical, economic, and environmental issues notwithstanding, desalination offers the Colorado River basin states an option for actually increasing water supplies. This option is limited primarily to areas with access to water derived from the Pacific Ocean, although there may be other, select Colorado River basin sites at which desalination facilities may be feasible (e.g., Yuma, Arizona). With increasing regional water demands, and with increases in technical efficiencies, desalination is likely to be perceived as an increasingly attractive option for augmenting supply, which will be especially true for wealthier communities. Prospective desalination projects will have to address and overcome the barriers of its energy requirements and acceptable means of disposing desalination's highly saline byproduct. Because of both its prospects and its potential limitations, desalination will also continue to be an important topic of research (see, e.g. *http://watercampws.uiuc.edu/*, which is a National Science Foundation-sponsored center for the study of materials and systems for safely and economically purifying water for human use).

REMOVING WATER-CONSUMING INVASIVE SPECIES

Since settlers began moving into the southwestern United States in the mid-19th century, many invasive species have been purposively or inadvertently introduced to the region. These include cheatgrass, camelthorn, ravenna grass, Russian olive, and tamarisk, or salt cedar. These species are identified by the National Park Service as the greatest threats to Grand Canyon National Park's native species (NPS, 2005) and are capable of surviving in a variety of habitats. They are today prolific within the park, representing approximately 10 percent of the vegetation (USGS, 2005).

Tamarisk (*Tamarix ramosissima*) is an invasive plant of special concern across the basin. Tamarisk consumes large quantities of water, crowds out native riparian species, and can lead to ecosystem-level changes. Tamarisk forms dense stands and is difficult to eradicate. Starting in the 1850s, several tamarisk species were imported to the United States as ornamentals and for use in erosion control (*http://www.invasivespeciesinfo.gov/council/ismonth/archives/tamarisk/tamarisk.html*). The plant eventually made its way to the Colorado River basin and spread upstream along the Colorado River to the lower end of Grand Canyon National Park in the 1930s. It did not take hold in Grand Canyon National Park or in the upper Colorado River basin, however, until large dams were constructed upstream in the 1960s (tamarisk had been previously controlled by large spring runoff from Colorado, Utah, and Wyoming flowing into the Grand Canyon). Tamarisk is a dominant riparian plant species today across the basin, consuming considerable amounts of water that would otherwise be available to downstream states or available to support ecosystem goods and services. For example, in Colorado it is estimated that tamarisk occupy roughly 55,000 acres and consume 170,000 acre-feet of water per year more than the native replaced vegetation (Colorado DNR, 2004). Smaller-scale efforts to control and remove tamarisk in the 1990s have led to more ambitious programs in the Colorado River basin and on other western rivers. A variety of efforts have been used to try to remove tamarisk, including herbicide injection, stump removal, deliberate flooding, and the use of a leaf beetle (*Diorhabda elongata*) and its larvae to eat tamarisk leaves. In 2002 the National Park Service, with support from environmental organizations and some 1,500 volunteers, began removing tamarisk from 63 tributary canyons along the Colorado within Grand Canyon National Park. To date, more than 180,000 tamarisk trees have been removed from 1,819 hectares in the park (NPS, 2006).

AGRICULTURAL WATER CONSERVATION

Agriculture water conservation in the western United States has long been an issue of widespread concern and importance. There are clearly inefficiencies in agricultural water applications in the West, some of which relate to difficulties in precisely matching water applications to plant water requirements (which will vary in response to changing temperature and soil moisture conditions). In addition, unused irrigation water on one farm usually generates return flow that is reused downstream by other irrigators and also helps sustain ecological habitats. Farm-level irrigation efficiency is of great importance to individual farmers; nevertheless, water withdrawals are "lost" from the stream reach from which they are withdrawn, which has ecological effects on instream flows and habitats.[1] Deep percolated water that finds its way to the groundwater table, however, may find its way back to a stream channel farther downstream. Farmers often periodically flush water through the soil profile in order to prevent excess salt accumulation in the crop root zone—especially if irrigation water is of marginal quality. This salt leaching practice is essential to successful irrigated agriculture (English et al., 2002). Leaching of salts from the root zone can, however, increase salt loading to streams and aquifers, which was one of the processes addressed in the Bureau of Reclamation's Colorado River Basin Salinity Control Program (which was mentioned in Chapter 2). There are opportunities for agriculture water conservation even in basins where water use efficiencies are high. Increased on-farm efficiencies can reduce production costs, improve water distribution among farmers and other users, and reduce negative, off-farm effects of irrigation (Wichelns, 2002).

One concept, although not new, is the idea of managing irrigation water as a means to maximize profit, not yield. Production of the final increment of crop yield requires disproportional amounts of water,

---

[1] From the perspective of the entire Colorado River basin, water that is reused downstream in the form of return flows is not lost, whereas water that evaporates from the soil, or that is transpired by phreatophytes or weeds, is lost to agriculture and other potential uses.

which can be relatively expensive. A strategy of not applying these additional units of water to gain this (relatively expensive) additional, final increment of production can save considerable amounts of water (Kelly and Ayers, 1982). There is also potential to employ different cropping patterns, such as rotations containing both sensitive and salt tolerant crops. It may be possible to grow a salt-sensitive crop, followed by a salt-tolerant crop, before salinity reaches unacceptable levels and soils must be leached (English et al., 2002; Manguerra and Garcia, 1996). Research on water use efficiency has been under way for years, including studies of water efficient technologies and ways of providing economic incentives and technical support for irrigators who adopt them (e.g., in the Grand Valley Unit with Reclamation's Salinity Control Program).

Irrigators will no doubt continue to innovate in managing their irrigation systems, taking advantage of remotely sensed soil moisture content and the state of crop growth—this latter approach includes deficit irrigation in which plants can be stressed at noncritical times but are properly watered at critical flowering and fruiting stages (English and Raja, 1996; Jurriens et al., 1996; Trimmer, 1990). Irrigation advisory services exist to provide assistance to farmers on these advanced techniques (English et al., 2002). In addition, some improvements in irrigation water productivity will likely result from better agronomy and cultivars, perhaps with genetic modification. There may also be prospects to increase water available for urban and instream uses by retiring some agricultural lands.

URBAN WATER CONSERVATION

As the Colorado River states have urbanized, growing water demands have stimulated more and more urban water conservation programs in many communities. These programs include water conserving technologies (e.g., lower-flow plumbing fixtures and more efficient irrigation systems), market incentives, regulatory policies, new landscaping techniques and the use of drought-tolerant (xerophytic) plants, and public education announcements encouraging urban water conservation. Elected officials today often promote water conservation plans to their constituencies and many citizens are willing to adopt household level water-saving practices. Many plumbing fixtures, such as toilets and shower heads, use less water than in a previous generation. This has given rise to periodic revisions of building codes and policies for retrofitting older commercial and residential structures. Use of reclaimed (or "gray") water for landscaping, golf course irrigation, and augmenting return flows to the Colorado River has also increased. The Southern Nevada Water Authority (SNWA), for example, reported the use of almost 22,000 acre-feet of reclaimed wastewater in its service area in 2003 (SNWA, 2004). A 1998 report from the NRC reviews the engineering, public health and policy issues associated with reclaimed water. That report focuses on implementing safe, potable standards for indirect uses, such as adding reclaimed water to other water supplies, then treating the mixed reclaimed and ambient water to conventional treatment standards. Although there are instances in which reclaimed wastewater represents a viable option, the report also identifies possible pitfalls in its applications (NRC, 1998).

Given rapidly growing urban water demands across the West, the impacts of increasing demands on reservoir storage levels and ecosystems, and the potential for urban water conservation and efficiency programs, there is widespread interest in approaches and technologies to help reduce urban water demands. Examples of recent studies, projects, and conferences in the West on this front include the following:

- The Bureau of Reclamation's Yield and Reliability Demonstrated in Xeriscape project. This study evaluated potential water savings and maintenance and installation costs associated with water-conserving landscapes. It was funded by the Bureau of Reclamation with in-kind services by seven participating municipalities and water districts in Colorado's Front Range. The final report was completed in December 2004 (Medina and Gumper, 2004).

- Industry-sponsored conservation studies and projects, such as those from the Aquacraft firm in Boulder, Colorado (*http://www.aquacraft.com/Services/water%20conservation.htm*).

- An American Water Works Association 2006 symposium in Albuquerque, New Mexico (*http://www.awwa.org/conferencesources/?CFID=14981257&CFTOKEN=69080727*).

- A series of water efficiency workshops and studies sponsored by Western Resource Advocates of Boulder, Colorado (*http://www.westernresourceadvocates.org/water/wateruse.php*).

Across the Colorado River region there are numerous approaches to managing and conserving urban water supplies. Some municipalities, such as Tucson, have aggressive and long-standing programs (see Box 4-3). The SNWA has instituted municipal and industrial conservation programs among the seven water and wastewater agencies that comprise its members (SNWA, 2004). Some of these have been quite useful; for example, SNWA estimates that Las Vegas area water use decreased by roughly 50,000 acre-feet per year between 2002 and 2005 because of implementation of a drought plan and a "Water Smart Landscape" program (Fulp, 2005a). Per capita water uses in the Colorado River basin's cities vary widely, from roughly 170 gallons per capita per day (gpcd) in Tucson to over 300 gpcd in some other cities in the basin (although because of the probability for comparison errors in this variable, these figures should be considered individually rather than as absolute comparisons; WRA, 2003). These differences reflect a large number of variables, including age of household water fixtures, conservation programs, municipal ordinances, water prices, and urban landscape expectations and norms.

They also suggest room for improvement in urban water conservation and efficiencies, and the value of disseminating lessons from successful urban water strategies in individual cities to other cities across the region. Knowledge of useful water conservation practices and techniques has diffused across the region through the efforts of groups such as the Colorado River Water Users Association. Nevertheless, there have been few efforts to systematically compare and evaluate the breadth and variety of these water conservation initiatives. One exception is a 2003 study from Western Resource Advocates that compares urban water uses across the southwestern United States (WRA, 2003). Efforts at comparing urban water management across the region, and sharing knowledge of successful experiences, could be enhanced by more formal water conservation program collaboration, and some mechanisms have been created to coordinate and support urban water programs (see, for example, the activities of the California Urban Water Conservation Council; *http://www.cuwcc.org/home.html*). Such efforts point to the prospects for improved regional urban water management and enhanced preparedness for managing water in periods of drought and water shortages.

OFFSTREAM WATER BANKING AND RESERVES

Water banking and groundwater recharge programs have been used for many decades in the western United States, and there has been an especially strong interest in these programs during the past decade. The term "water bank" generally applies to two different types of arrangements: (1) groundwater storage projects, and (2) arrangements to facilitate voluntary water transfers through rental markets (http://www.isse.ucar.edu/water_climate/banking.html). Water banking and groundwater storage programs have several objectives, which include (1) the (hoped-for) creation of reliable supplies during extended drought, (2) the promotion of water conservation by encouraging "deposits" into groundwater storage, and (3) the recharging of groundwater tables and reduction of evaporation from surface reservoirs. From a geological perspective, large amounts of water can often be infiltrated, via gravity, into under ground aquifers in many locations. However, during drought conditions, large amounts of water may need to be withdrawn in a very short period of time, which often entails significant pumping costs. Groundwater storage programs aim to facilitate water transfers in response to short-term changes in supply-and-demand conditions, with the goal to bring together people seeking to purchase water with people interested in selling water entitlements (Frederick, 2001; MacDonnell et al., 1995). Salient featuresof recent initiatives involving banking and exchanging Colorado River basin water are discussed below.

The State of Arizona created its first framework for water banking in 1986, with passage of legislation to authorize underground storage and recovery projects. In 1996 the Arizona Water Banking Authority (AWBA) was established. The AWBA focused on storing surplus Colorado River water through groundwater recharge and on protecting Arizona's Colorado River supplies by demonstrating the state's commitment to using its full allocation. This groundwater supply could subsequently be drawn upon for use during shortages of Colorado River flow, during Central Arizona Project service disruptions, to assist in meeting management objectives of the Arizona Groundwa-

> **BOX 4-3**
> **Water Conservation in Tucson**
>
> Water conservation programs for the City of Tucson, Arizona, offer an interesting case study for several reasons. First, some of the programs date back to the 1970s and are among the oldest urban water conservation efforts in the region. Second, these programs have included several different aspects including public education, water-saving technologies, water pricing, and regulation. Third, the Tucson Water Department has maintained an excellent database on its conservation programs and now, after four decades, has a compendium of valuable information. Finally, Tucson has constantly revised its policies and strategies to incorporate newer, efficient technologies, updated water pricing and revised regulations, for example, codes for low-water-use landscaping (xeriscaping) and drip irrigation systems.
>
> In 1973 Tucson annual per capita water use reached an all-time high of 205 gallons per day. The city was unprepared to ensure reliable service, a warm summer in 1974 led to increased water use rates, and both household and industrial growth were accelerating. Although reliability of service was largely a summertime peak demand issue, the city instituted year-round water conservation policies. Through a series of measures, water consumption dropped to roughly 150 gallons per capita/day, counting both household and industrial users.
>
> Tucson's urban water conservation program is built around five interrelated strategies:

ter Code, and to assist in meeting Native American water rights claims settlements. The AWBA also provides some insurance to offset liabilities associated with the Central Arizona Project's junior water right, which is subordinate to California's 4.4 million acre-feet per year Colorado River allocation. By 2000 the AWBA was recharging about 294,000 acre feet per year of water (*http://www.awba. state.az.us/backgrnd/update.html*). The practice allows storage of portions of the state's allotment that are not utilized at present and storage of surplus water during years of high river flow. Water is recharged into suitable geologic basins in western and southern Arizona, where it is not susceptible to evaporation and where it can be accessed with relatively modest pumping costs.

In 1999 the Secretary of the Interior published regulations defining the procedure for the lower basin states to engage in interstate offstream storage agreements (see 43 CFR 414.3). These regulations

> - general public information,
> - education and training programs,
> - incentive programs,
> - direct assistance, and
> - regulatory measures.
>
> The City of Tucson began delivering reclaimed water in the mid-1980s; a large percentage of parks, golf courses, and other public spaces today are irrigated with reclaimed water. Beyond the 1980s, domestic household usage rose to about 170 gallons per capita per day (gpcd) and has since been relatively stable. Xeriscaping is mandated by building code. Tucson maintains data for analysis (e.g. gpcd water consumption in single-family and multi-family dwellings) and water use trends are studied. Tucson's programs have been complemented by state legislation. For example, the Arizona Department of Water Resources code established standards for the reduction of per capita water use. Tucson's water plan since 1990 has included periodic assessment of the effectiveness of programs, development of ways to continue promoting the five-part conservation strategy, and infrastructure improvements that capture lost water (e.g., replacement water pipelines to reduce water losses through leaks). Today Tucson is looking to a new conservation plan to guide water uses over the next two decades (more information on Tucson Water is available at *http://www.ci.tucson.az.us/water/*).

set the groundwork for a subsequent interstate water banking agreement (a storage and interstate release agreement, or SIRA) among the AWBA, the Colorado River Commission of Nevada, and the SNWA. A 2004 amendment to an existing agreement with AWBA allows the Nevada water agencies access to 1.25 million acre-feet of water in the Arizona Water Bank. Banked water is stored in the form of "credits." For Nevada to recover its storage credits, Arizona will use banked water and forego the credited amount of Colorado River water to Nevada. The Nevada agencies will then divert water from Lake Mead (Davenport, 2005; SNWA, 2006). In 2004 the Metropolitan Water District of Southern California entered into a similar, but smaller scale, water banking agreement with the SNWA and the Colorado River Commission of Nevada (Davenport, 2005) that establishes terms and conditions for offstream storage of Colorado River water in Southern California and provides storage credits for SNWA (DOI and USBR, 2006). In the future, these and other innovative types of interstate groundwater storage and banking initiatives are likely to be implemented elsewhere in the Colorado River basin where there are willing parties, where requisite geologic and other physical conditions exist, and where necessary institutional changes can be made.

A related concept being explored in the basin involves the creation of water reserves, which is not to be confused with the concept of reserved rights that can exist under the doctrine of prior appropriation (Box 4-4 discusses aspects of a recent water reserve proposal in the State of New Mexico). The concept of water reserves generally entails the storage of water, either by excess flows in wet periods or via water rights sales, leases, or transfers, to be used at a later date for a specific purpose(s). It is a variant on the water banking concept in that it is not necessarily fully market-based and may be designed to benefit public and nonmarket values (e.g., instream flows).

Box 4-4
New Mexico's Strategic Water Reserve

A proposal to create a strategic river reserve in the State of New Mexico was offered in 2003. The proposal was drafted by Think New Mexico, a research institute that sought to raise awareness regarding the state's fully allocated and overallocated rivers, the high cost of prior water litigation in the state and with neighboring states, water needs for agriculture and environmental purposes (e.g. endangered species in the middle portion of the Rio Grande), and increasing demands that population growth was placing on the state's five major rivers. Think New Mexico recommended that the legislature enact and fund a strategic river reserve. The concept was endorsed in the legislature and was promoted through editorials and community-based support. The New Mexico governor called for the reserve in his 2005 state-of-the-state address. A strategic water reserve bill was enacted in the first session of the 47th legislature in 2005 and funded in the 2006 second session in the amount of $4.8 million. The fund is administered by the New Mexico Interstate Stream Commission, which will purchase or lease water rights that become available and redeploy them for public purposes including agriculture, endangered species, and assurance that the state meets its obligations for interstate water transfers, especially downstream in the Rio Grande and Pecos rivers to Texas.

SOURCE: Think New Mexico (2003); NM Statute 72-14-3.3 et seq.

COMMENTARY

There has been a wide range of engineering and political efforts designed to overcome the water supply limitations imposed by the aridity westward of the 100th meridian. There are technological means available to extend water supplies in the Colorado River basin and elsewhere, but all these options have limits. Although limited opportunities exist to construct additional reservoirs or to implement interbasin water transfers into the Colorado River basin, these have diminished from a previous era. Changing economics and demographic conditions may increase the viability of such traditional projects at some point in the future, but immediate prospects for major new water supply reservoirs or interbasin transfers are limited. Consequently, new water project prototypes that emphasize conservation, landscaping, new technologies, and other measures are being promoted across the West.

Desalination certainly represents an alternative for augmenting water supplies in some circumstances but it can be expensive and may not always be feasible. Disposal of brine water can be problematic, for example, and in many instances desalination is a realistic option only for coastal cities. Cloud seeding may offer marginal opportunities for increasing supply—especially in the upper basin states—but it does not appear to offer a reliable long-term means for increasing precipitation and water supplies. Groundwater banking and offstream water reserve programs have proven useful in many instances and are being used in more areas and instances but they are limited by geologic conditions. Agricultural-urban water transfers are also likely to be effected more often in the future. As noted in Chapter 2, such transfers represent lucrative opportunities for both buyers and sellers and often entail third-party effects, all of which are important to consider when negotiating water leasing and transfer arrangements.

Given the projections of both increasing regional population and increasing regional temperatures, along with tree-ring-based reconstructions that demonstrate the recurrence of severe drought conditions across the Colorado River region, urban water conservation will become only more important. Future augmentation of urban water supplies can, and will, be achieved through a variety of water conservation, pricing, and other measures. Water consumption and conservation practices are strongly related to water prices, incentives, and regulations. If water prices markedly increase, people and businesses will use less water; however, water prices have always been tied closely to political decisions and this is not likely to change in the near future. Incentives can help reduce per capita water use, as can tighter regulations and fines for excessive water use. There have been many studies and reports regarding what might be accomplished through nonstructural measures designed to conserve water (e.g., Gleick et al., 2005). Clearly, there are gains to be realized through aggressive water conservation measures. There is no formal basin-wide strategy or program designed to promote urban water conservation across all cities. There are, however, programs such as the California Urban Water Conservation Council that supports statewide urban water use programs, and other, similar efforts could lead to further water use efficiencies. But broadly speaking, none of the technological or strategic options for either increasing or conserving and extending water supplies examined in this chapter directly confronts the relationships between urban population growth, water demands, and limited water supplies in this arid region.

Technological and conservation options for augmenting or extending water supplies—although useful and necessary—in the long run will not constitute a panacea for coping with the reality that water supplies in the Colorado River basin are limited and that demand is inexorably rising.

Proper water management under normal climate and hydrologic conditions poses many challenges, and under drought conditions, such challenges are greatly magnified. This is an especially important concern given regional warming trends and long-term climate studies indicating that long-term droughts recur periodically across the Colorado River basin. Chapter 5 lists some important issues in adjusting to drought and identifies and discusses some of the key organizations and programs focused on improving drought preparedness in the Colorado River region.

5

Colorado River Basin Drought Planning Strategies and Organizations

Water managers in the Colorado River basin and across the U.S. West have long coped with occasional drought and water shortages. The drought of the early 2000s will surely be followed at some point by wetter conditions. Nevertheless, findings derived from tree-ring-based reconstructions, along with temperature trends and projections for the western United States, point toward a future in which droughts—some of them severe—are likely to recur with greater frequency and duration. And, as explained by the U.S. Geological Survey in describing possible impacts of drought conditions such as those experienced from the 1940s though the late 1970s across the region, "The region's population has increased fourfold since the mid-1950s, creating the possibility of severe consequences if such a drought were repeated" (USGS, 2002). Increasing urban populations and water demands over the past two decades have highlighted the importance of drought planning in the context of urban water management. The value of preparing for, detecting, and responding to Colorado River drought conditions will only become greater in the future.

Drought is a recurrent phenomenon across the western United States; in fact, historical records show that drought occurs somewhere in the West almost every year (Wilhite, 1997a). Droughts are part of normal climate patterns in the Colorado River region but they do not occur with any clearly identified regularity and are difficult to forecast (see Box 2-3). Drought is a slow-onset event and drought conditions are often well under way before its presence is widely recognized. Moreover, the lack of universal standards for defining drought means that it is not always clear when drought has begun or ended.

Effective drought planning may be best accomplished in periods of water surplus, but there are often few compelling incentives to develop drought management plans during periods of high precipitation and "surplus" water. Human nature being what it is, droughts are not easy to anticipate and carefully plan for.

For much of the 20th century the traditional approach for coping with periodic water shortages (and to spur development) in the Colorado River basin was to construct storage reservoirs with sufficient capacity to both support future growth and meet water demands during drought. This practice was viable for many years; however, this strategy requires access to untapped (and previously undammed) water sources, good reservoir sites, strong congressional support, and a citizenry willing to accept the environmental costs associated with dam construction. Not all these conditions exist as they did in the early and mid-20th century and the prospects for expanding storage capacity via large federal reservoirs are essentially at an end across the West. And, regardless of human desires, dams do not create water resources, they only allow storage of natural precipitation and streamflow. During the 1950s and 1960s, water storage capacity greatly expanded in the Colorado River basin, with Lake Mead and Lake Powell providing a combined storage capacity of roughly 55 million acre-feet—almost four times the river's annual average flow. As a result, water supply problems afflicting the Colorado River basin today thus relate less to storage capacity (for example, in 2006 most of the basin's reservoirs were well below capacity and could store much more water) and more to limited supply, as well as to incessant increases in water demands. Moreover, as former Bureau of Reclamation Commissioner John Keys stated, "The days of the large [dam] project with the federal government as the sole funding source are over" (Keys, 2006). This of course does not mean there will be no new water storage projects, but it does mean that new projects will require close cooperation between water users, and multiple parties to plan, finance, build, operate, and maintain facilities. This report notes that water projects of the future are less likely to entail new dams and reservoirs and more likely will be focused on urban water conservation, landscaping, education programs, and better management of existing supplies. Accordingly, there will likely be some shifts in how organizations and citizens cope with recurrent drought and water shortages.

Regardless of when drought conditions might abate, the Colorado River basin states face some sobering prospects with regard to the balance between long-term water availability and demands. Increasing population growth and water demands mean that the Colorado River storage system will have less water available in storage and will take longer to recover in future droughts (Fulp, 2005a). Good drought detection, mitigation, and preparedness programs thus will be increasingly important. This chapter discusses issues of drought planning and coping with water shortages, especially in the rapidly urbanizing Colorado River basin. It identifies organizations, programs, and studies aimed at improving drought planning and response.[1] It also discusses strategies that municipalities have used to help conserve water, especially during drought.

FEDERAL-LEVEL PROGRAMS

Drought Policy Legislation and the National Integrated Drought Information System

The 1990s and early 2000s saw substantial efforts to enhance national-level drought preparedness, through federal legislation and through the creation of a National Integrated Drought Information System (NIDIS). The Western Governors Association (WGA) provided a strong impetus for these initiatives. In 1996, the western governors set a goal to change the way the nation prepares for and responds to droughts, calling for a national policy to be enacted "which provides for a comprehensive, coordinated and integrated approach to future droughts" (see *http://www.westgov.org/wga/initiatives/drought2.htm*). The western governors soon thereafter adopted the Drought Response Plan of 1996, which included recommendations to improve federal and state responses to droughts. The plan also called for the development of a national drought policy or framework to integrate federal, state, regional, and local actions. With strong support from WGA, the National Drought Policy Act of 1998 was signed into law. That 1998 act established the National Drought Policy Commis-

---

[1] See *http://wwa.colorado.edu/resources/colorado_river/management_use.html* for a more extensive listing of some of these entities.

sion, which issued a report in 2000 that served as a basis for future drought-specific legislation. Then, in 2003, the National Drought Preparedness Act was first introduced in the U.S. Senate, with a companion bill introduced in the U.S. House of Representatives (different versions of that draft legislation currently are pending in Congress). In 2006, the NIDIS Act was introduced to enhance the nation's drought early warning system, provide drought monitoring, and develop drought policy and planning techniques. The NIDIS is likely to be housed within the National Oceanic and Atmospheric Administration and will collaborate with partners such as the National Drought Mitigation Center at the University of Nebraska. As this report went to press, that bill had passed the U.S. House of Representatives and was awaiting consideration by the Senate.

Bureau of Reclamation

The U.S. Bureau of Reclamation is one of three primary federal agencies with drought-related responsibilities (the other two are the U.S. Department of Agriculture and the U.S. Army Corps of Engineers). Title II of the Reclamation States Emergency Drought Relief Act of 1991 (P.L. 102-250) authorizes the Bureau of Reclamation to undertake drought mitigation activities in consultation with other appropriate federal and state officials (of all 50 states and U.S. territories); tribes; and public, private, and local entities. The two major components of the program (1) relate to response activities during times of actual drought events for construction of temporary facilities, management, and conservation measures to minimize drought-related losses, and (2) provide assistance in preparing plans to prevent and mitigate effects from future drought events. This is usually contingent upon annual federal appropriations, frequently through an emergency supplemental bill (as opposed to sustained annual funding).

The Bureau of Reclamation is involved in several drought-specific programs that provide technical assistance to state and local agencies (e.g., irrigation districts) regarding water management alternatives and system improvements. Examples of these activities include assisting project offices in forecasting water supplies, assisting in water transfers requiring use of Bureau of Reclamation project facilities, modifying project facilities or operations, and helping develop state drought indices. In managing its water storage and delivery system across the Colorado River basin, the Bureau of Reclamation considers ways in which it might help mitigate drought conditions; as this report went to press, for example, the Bureau of Reclamation released a draft environmental impact statement regarding coordinated operations for Lakes Powell and Mead under low reservoir conditions (*http://www.usbr.gov/lc/region/programs/strategies/draftEIS/index.html*). As explained in Chapter 2, the Bureau of Reclamation is responsible for implementing many provisions of the Colorado River Compact and the Law of the River, such as water delivery obligations from the upper basin to the lower basin, and from the United States to Mexico. In addition to legal obligations, the Bureau of Reclamation makes many operations decisions in an effort to balance different, often competing, management objectives. Reclamation employs real time data, weather and climate forecasts, and water demand forecasts as inputs to decision support systems used in systems operations. The Bureau of Reclamation makes these decisions on a variety of time scales, as reflected in documents such as its Annual Operating Plan.

National Oceanic and Atmospheric Administration

Colorado Basin River Forecast Center

The Colorado Basin River Forecast Center (CBRFC) is part of the National Weather Service and is located in Salt Lake City. The CBRFC issues operational forecasts for upper and lower Colorado River basin weather (along with local offices) and streamflow, on time scales ranging from minutes to seasons. The center provides forecasts of April through July runoff to the Bureau of Reclamation as the year progresses. Streamflow forecasts in the headwaters regions are issued jointly by the Natural Resources Conservation Service Water and Climate Center (located in Portland, Oregon) and by CBRFC. Forecasts of streamflow on the Colorado River mainstem are primarily developed by CBRFC, which maintains a technical database of river systems across the region.

Regional Integrated Sciences and Assessment

The Regional Integrated Sciences and Assessment (RISA) program combines scientific expertise from government and academic institutes to support research that addresses climate-related issues of concern to policy planners and decision makers at a regional level. Colorado River region RISA programs include the Western Water Assessment at the University of Colorado, the Climate Assessment of the Southwest at the University of Arizona, and the California Applications Project at Scripps Institution of Oceanography (*http://www.climate.noaa.gov/cpo_pa/risa/*). All these programs contain research foci concerning climate variability and change, in concert with water supplies and impacts on socioeconomic sectors. The RISA programs work with water managers to assist them in using climate information in decision making.

U.S. Department of Agriculture

The U.S. Department of Agriculture (USDA) and its Natural Resources Conservation Service (NRCS) are involved in several drought-related programs. USDA provides a variety of drought assistance programs, including disaster assistance and emergency loans. The NRCS promotes drought awareness and preparation through its "Defending against Drought" program (see *http://www.nrcs.usda.gov/feature/highlights/drought.html*). NRCS also sponsors the National Water and Climate Center, which provides information on snowpack and water supply forecasts in the western United States.

STATE-LEVEL PROGRAMS

Drought is usually experienced initially at the local and regional levels, and given the limited history of national-level programs to address drought, states have emerged as important innovators in ways to reduced long-term vulnerability to drought (*http://www.drought.unl.edu/mitigate/status.htm*). A 1997 paper prepared for the Western Water Policy Review Advisory Commission noted that, "In the United States, States are clearly the policy innovators for drought management" (Wilhite, 1997b). Much of the innovation on the drought preparedness and planning front has taken place in the past three decades. During the U.S. drought of 1976–1977, no state had a formal drought plan, and in 1982 only three states had drought plans. But as of October 2006, 37 states had drought plans, 2 states delegated planning to local authorities instead of having a single state-level plan, and 2 states were in the process of developing a plan. Only nine states today do not have formal drought plans (Wilhite, 1997b). All the Colorado River basin states have some type of formal drought action or management plans, with varying emphases on mitigation, response, and delegating drought planning to local entities. In fact, one of the first states to develop a drought plan was Colorado (Wilhite, 1997b). When Colorado developed the plan in 1981 it was one of the three states in the nation with a drought plan. The plan has since been revised in order to improve the state's capacity to cope with water shortages. The Colorado Drought Mitigation and Response Plan is administered by the Office of Emergency Management under the authority of the Colorado Department of Natural Resources (*http://cwcb.state.co.us/Conservation/pdfsDocs/ColoradoDroughtResponsePlan.pdf*). More information, including an up-to-date and comprehensive listing of the various state-level drought planning programs in the Colorado River basin (and the entire nation) is available through the National Drought Mitigation Center (*http://drought.unl.edu/plan/ stateplans.htm;*).

Western Governors Association and the Western States Water Council

The Western States Water Council (WSWC) is headquartered in Midvale, Utah and was created by the WGA in 1965. Its purposes are to (1) accomplish effective cooperation among western states in conservation, development, and management of water resources; (2) maintain state prerogatives while identifying ways to accommodate legitimate federal interests; (3) provide a forum for the exchange of views, perspectives, and experiences among member states; and (4) provide analysis of federal and state developments in order to assist member states in evaluating impacts of federal laws and programs and the effectiveness of state laws and policies. The 17 western contiguous states and Alaska are members of the WSWC, which has a full-time staff at its Midvale headquarters and issues reports on many different western water policy issues, including drought (see WGA

[2006] for a WSWC-WGA report on water needs and strategies; see WGA [2004] for a report on drought response). The WSWC has also been instrumental in promoting federal legislation and related actions aimed at drought mitigation.

The WGA has played a pivotal role in western drought activities since the Texas-Oklahoma-New Mexico drought of 1995-1996, which led to the formation of the Western Drought Coordinating Council (WDCC) in 1997. Since then, drought has been present in each successive year through 2006 somewhere in the 11 westernmost states. The WDCC led to the formation of the National Drought Policy Commission in 1999 and the Interim National Drought Council in 2000. These groups stimulated several more activities of relevance to the Colorado River at the federal level, in which the WGA played a major role, including development of the NIDIS.

Interstate Cooperation on Colorado River Water Shortages

An issue strongly related to drought planning and mitigation across the Colorado River basin is depleted storage levels and Colorado River flows. The Colorado River Compact and its provisions for allocating river flows among states represent a management framework and philosophy with numerous interstate ramifications. Since Lake Powell initially filled in the early 1980s, there had always been ample water in the lake to meet water release obligations to the lower basin, and the basin states had never developed shortage guidelines. Since then, however, population and water demands in the basin states have steadily increased and have put additional demands and pressures on Lake Powell and other water storage facilities. The Colorado River water supply-and-demand dynamic was changing rapidly and led to an interesting paradox as drought conditions deepened in the early 2000s. For example, at that time there were still concerns over how the basin states might share "surplus" waters in Colorado River reservoirs and, in January 2001, then-Secretary of the Interior Babbitt approved a set of rules known as Interim Surplus Guidelines (Garrick and Jacobs, 2006). But very quickly, the drought of the early 2000s brought the issue of interstate cooperation on coping with Colorado River water shortages to a head, and "[f]ew imagined the transition from surplus to shortage would occur so soon"

(Garrick and Jacobs, 2006). As the drought of the early 2000s reached full swing the states began discussing how they might better cooperate in coping with Colorado River water shortages.

The drought prompted the basin states to request the Secretary of the Interior to operate Lake Powell and Lake Mead differently; namely, the upper basin states requested that Lake Powell releases be reduced from the traditional minimum of 8.23 million acre-feet per year if the drought continued. The Secretary countered, challenging the basin states "to work together and present to her an alternative acceptable to all seven states for her to include in the EIS she had instructed the Bureau of Reclamation to prepare" (Anderson, 2006). In a notable development, the seven basin states, via a February 3, 2006 letter to the Secretary of the Interior (Appendix A), developed a preliminary shortage management proposal. The proposal, developed in cooperation with the Bureau of Reclamation and its long-range planning model—the Colorado River Simulation System—attempts to balance competing demands within the existing Law of the River framework (Garrick and Jacobs, 2006). The actual decision regarding shortage guidelines is not final and is subject to an environmental impact statement that is considering how to address the issue of limited water availability during times of low reservoir conditions (the Bureau of Reclamation released its draft EIS in February 2007; see *http://www.usbr.gov/lc/region/programs/strategies/draftEIS/index.html*). The cooperation reflected by the February 2006 letter from the basin states—which includes agreements on the coordinated management of Lake Powell and Lake Mead, along with other specific provisions—will be an increasingly important part of viable drought preparedness strategies.

MUNICIPAL-LEVEL PROGRAMS

Several municipalities that use Colorado River water have drought education, preparedness, and response programs, many of which have been successful in reducing urban water demands and in increasing water use efficiencies. In California, for example, the Metropolitan Water District (MWD) of Southern California is a consortium of cities and water districts that provides drinking water to nearly 18 million people in parts of Los Angeles, Orange, San Diego, Riverside, San Bernardino, and Ventura counties. MWD has several initiatives aimed at water conservation and recycling. Its water conservation programs are guided by MWD's "Integrated Resources Plan" and by the California Urban Water Conservation Council's Memorandum of Understanding Regarding Urban Water Conservation in California.

In Colorado, Denver Water provides water to over 1 million people in the Denver metropolitan region and surrounding communities and water districts (Kenney et al., 2004). Denver Water has an extensive water conservation program that includes xeriscaping education and assistance, advice on self-audits, irrigation efficiency, and other water saving measures. These programs have good potential for helping conserve water supplies and proved effective at reducing water uses in the early 2000s. Through a combination of measures, Denver Water and several other water providers in Colorado's Front Range were able to reduce per capita water uses in 2002. Periods of mandatory water restrictions were especially effective, resulting in per capita savings ranging from 18 to 56 percent, as compared to 4-12 percent savings during periods of voluntary restrictions (Kenney et al., 2004).

In Nevada, the Southern Nevada Water Authority (SNWA) is a cooperative agency governed by seven water districts and municipalities in the region, including the cities of Boulder City, Henderson, and Las Vegas. It relies heavily on the Colorado River, from which it derives 90 percent of its water. SNWA promotes water conservation and efficiency through water restrictions, water savings rebates, and several other programs (e.g., xeriphytic landscaping and modern irrigation technologies). SNWA also issued a Drought Plan in 2005, which was developed initially in response to the drought across the region in the early 2000s. The plan outlines water demands, conservation goals, and drought response measures. SNWA's efforts have paid off, as Southern Nevada's consumptive water use declined by about 20 billion gallons between 2002 and 2005, despite the fact that the region added nearly 250,000 new residents in this same period (*http://www.snwa. com/html/drought_index.html*).

Although there have been many innovative urban water conservation programs and strategies across the Colorado River region in the past decade, no organization or program formally documents or otherwise coordinates these various urban water conservation measures, regional water forecasting techniques, or drought planning strategies.

There are also few efforts to compare and contrast the many water conservation activities at the municipal or household levels, or to compare historical strategies and initiatives for coping with drought conditions, across the region. Several organizations and studies, however, have promoted drought preparedness activities and better management of urban and other water resources during drought periods.

OTHER ORGANIZATIONS AND INITIATIVES

The National Drought Mitigation Center

The National Drought Mitigation Center (NDMC), located at the University of Nebraska-Lincoln, helps people and institutions develop and implement measures to reduce societal vulnerability to drought. The center stresses drought preparedness and risk management (as opposed to crisis management) in coping with drought. Most of NDMC's services are directed to state, federal, regional, and tribal governments that are involved in drought and water supply planning. Its primary activities include maintaining an information clearinghouse and drought portal; drought monitoring, including participation in the preparation of the U.S. "Drought Monitor" (see below) and maintenance of the Drought Monitor website (*http://www.drought. unl.edu/dm/monitor.html*); drought planning and mitigation; drought policy; advising policy makers; collaborative research; K-12 outreach; workshops for federal, state, and foreign governments and international organizations; organizing and conducting seminars, workshops, and conferences; and providing data to and answering questions for the media and the general public. The NDMC also participates in international projects, including establishment of regional drought preparedness networks in collaboration with the United Nations' Secretariat for the International Strategy for Disaster Reduction.

In response to a series of droughts in the West and a developing drought in the northeastern United States, the Drought Monitor was first posted on the Internet on May 20, 1999. The Drought Monitor is both a process and a product. The process involves electronic receipt of weekly input from about 150-200 federal, regional, state, and university drought specialists offering information on climate and on impacts in their geographical or topical area of expertise. Drought Monitor participants are from the NDMC, the National Oceanic and Atmospheric Administration, and USDA. These groups synthesize information on drought and its impacts and determine a weekly classification at every location in the country. This is vetted in an iterative national exchange, with the product being the Drought Monitor map posted at the NDMC website. By all accounts this experiment has been successful in generating discussions that have stimulated several operational products and important research questions (see a series of articles in a 2002 issue of the Bulletin of the American Meteorological Society: Heim, 2002; Keyantash and Dracup, 2002; Redmond, 2002; Svoboda et al., 2002).

Urban Water Management and Collaboration

Many nonprofit organizations and recent studies focus on the issues of drought, water shortages, urban water management, climate change and variability, and the links between these topics. It would require an extensive effort and resources to identify and describe all the groups and experts involved in these studies, but this section provides select examples of these initiatives. In doing so, it demonstrates that organizations, experts, and citizens in the Colorado River states are increasingly realizing the importance of urban water management, population growth, and limited water supplies.

The Colorado River Water Users Association (CRWUA) was established to enhance personal relations and communications among water agencies from the seven Colorado River basin states (see also *http://www.crwua.org/*). It convenes a well-attended annual conference in Las Vegas, with a timely theme and a variety of presentations and poster sessions. CRWUA and its annual conferences have surely enhanced communication among water users and officials across the region. The CRWUA considers the full spectrum of water-related issues of relevance and interest to its members. On the urban water front, the California Urban Water Conservation Council (CUWCC), headquartered in Sacramento, was created to increase efficient water use statewide through partnerships among urban water agencies, public interest organizations, and private entities. Its goal is to integrate urban water conservation best management practices into the planning and management of California's water resources. A memorandum of understanding was signed by nearly 100 water agencies in 1991; since then, CUWCC has grown to 354 members (see *http://www.cuwcc.org/home.html*). Groups such as CRWUA and CUWCC have contributed greatly to sharing information on urban and other water practices among professionals.

Studies and Workshops on Drought, Climate Change, and Urban Water Management

The issues of drought monitoring, preparedness, impacts, and response are frequent topics at workshops and meetings, with a variety of sponsors and participants, across the western United States. The interest surrounding drought impacts in the western United States and Colorado River basin is not new (Box 5-1, for example, describes a major drought study conducted in the early and mid-1990s). There is no question, though, that the drought of the early 2000s sparked intense interest across the region in drought-related topics.

The drought of the early 2000s and its implications for the Colorado River Compact were the focus of the University of Colorado Natural Resources Law Center Annual Conference in July 2005 (*http://wwa.colorado.edu/resources/colorado_river/hard_times_conference/index.html*). The New Mexico School of Law Water Policy Conference featured discussion of federal and state institutional responses to drought at its annual conference in May 2005. And in September 2006 a meeting of the NIDIS was convened in Longmont, Colorado, with participants from federal and state agencies, the NDMC, and the WGA. Other examples of recent studies on urban water and water shortages across the region include a 2003 study from Western Resources Advocates in Boulder, Colorado (WRA, 2003) that compares urban water efficiencies across the Colorado River states; a 2006 report on the linkages between water utility operations and drought and climate change (e.g., AWWA and UCAR, 2006); and a 2006 report on Arizona water management innovations,

> **BOX 5-1**
> **Colorado River Basin Severe Sustained Drought Study**
>
> In the early 1990s the U.S. Army Corps of Engineers and the U.S. Department of the Interior provided funding for a study on the effects of a major drought on the Colorado River basin. The Severe Sustained Drought (SSD) study was overseen by the Powell Consortium and was conducted by an interdisciplinary team of water resources experts from across the region. It included studies in several topical areas, including tree-ring reconstructions of historic runoff, hydrologic analyses of the probability distribution of river flows, engineering simulations of the functioning of water management facilities and institutions, and legal and institutional analyses of interstate water allocation rules (Lord et al., 1995). The study lasted roughly 10 years and one of its primary products was a series of papers published in a 1995 edition of the *Water Resources Bulletin*, the journal of the American Water Resources Association. The SSD study led to several interesting findings and was by all indications a useful exercise in thinking about long-term impacts of, and system responses to, severe drought.
>
> One interesting point regarding the SSD exercise was the construction of a severe drought scenario. SSD participants wanted to consider extreme drought conditions, and consulted the long-term tree-ring record in a search for severe conditions. Consultant Ben Harding of Hydrosphere Resource Consultants in Boulder, Colorado, worked with the SSD team to create a drought of unprecedented severity, "just the worst kind of drought you could possibly contemplate" (quoted in Jenkins, 2005). Drought conditions in the early 21st century, however, exceeded even this worst-case scenario used in the SSD, in part "because . . . water uses in the Lower Basin are higher (than was modeled)" (Jenkins, 2005).

with an emphasis on managing scarce water supplies in the face of rapid urban growth (Colby and Jacobs, 2006).

Organizations such as the CRWUA and CUWCC generally do not have the mandate or resources to gather and systematically evaluate information and issue reports on water use and conservation experiences across the region. There have also been few studies aimed at broad comparisons of urban water management across the entire Colorado River basin (the 2003 study from Western Resource Advocates represents an exception). As discussed above, most drought-planning across the region has tended to be at the municipal and state level; there are signs of increasing cooperation on this front in the form of initiatives like the National Drought Policy Commission and NIDIS. Knowledge of successful and innovative programs for managing urban water during shortage periods tends to be anecdotal, reducing the chances that water managers will benefit by learning of experiences from across the region.

A systematic project or study to document and synthesize urban water use strategies from across the region would be a useful reference for municipalities in the Colorado River region, could further encourage interstate cooperation on drought planning, and could provide useful information to other parts of the nation that are experiencing increased water demands and are challenged to meet water demands during periods of drought—especially since it is increasingly appreciated that drought and water shortages are not limited to the arid western states.

COMMENTARY

The drought of the early 2000s placed heavy demands on the Colorado River basin storage system. Despite sharp drops in storage levels in many of the system's primary reservoirs, the system demonstrated significant capacity to cope with extended drought conditions and in many respects performed as designed. Whether the system could have adequately handled a longer or more severe drought, however, is an open question and one that should cause water managers and elected officials to consider the basin's capacity to cope with severe, long-term droughts. This is an especially important issue given the collective evidence from tree-ring-based reconstructions of Colorado River flows, trends and projections that reflect increasing temperatures, and rapidly growing population and urban water demands.

Drought conditions tested the region's institutional capacity to cope with water shortages and gave rise to positive developments in terms of interstate cooperation, scientific information exchange, and a heightened awareness that—even without future severe drought—increasing water demands will continue to stress water supplies. The February 2006 letter from the seven basin states to the Secretary of the Interior that approved a preliminary shortage management proposal bodes well for future cooperation. It will surely be something that the states can build upon in future interstate negotiations regarding drought and water shortages.

The interstate cooperation and initiative exhibited by the Colorado River basin states in their February 2006 letter to the Secretary of the Interior is a welcome development that will prove increasingly valuable—and likely essential—in coping with future droughts and growing water demands.

Several developments during the drought of the late 1990s and early 2000s promoted communication among the climate sciences community and Colorado River water managers. Many conferences and workshops on drought and water availability were convened and several federal-level initiatives—such as the 2000 report from the National Drought Policy Commission, and support for development of NIDIS—point to a greater emphasis on drought preparedness and communication among climate scientists and the water management community. Lines of communication that were opened and strengthened among the climate science community and Colorado River water managers during the early 2000s drought represented a welcome development. The hydroclimatic sciences community studying drought and water resources across the Colorado River region is large and diverse, encompassing many research topics and themes. It may be impractical for elected officials and federal, state, and municipal water system administrators to follow all relevant developments in these scientific fields; but periodic discussions among climate science experts and water managers can help water system decision makers stay abreast of recent developments. It can also allow water managers to help frame scientific questions and lines of inquiry that would be useful for the water management community. Strong and sustained two-way dialogue will also help climate scientists better understand the legal and political context of decision making, budgets of administrative units, and issues of concern to the water management community.

A commitment to two-way communication among scientists and water managers is important and necessary in improving overall preparedness and planning for drought and other water shortages. Active communication among people in these communities should become a permanent fixture within the basin, irrespective of water conditions at any given time. Such dialogue should help scientists frame their investigations toward questions and topics of importance to water managers, and should help water managers keep abreast of recent scientific developments and findings.

The Colorado River Compact and much of the Law of the River were framed during an era in which water for irrigation was of paramount concern. This emphasis on Colorado River water for irrigated agriculture has been in a state of flux for many years. Environmental concerns caused a shift in Colorado River management priorities beginning in the 1970s, and marked increases in urban population growth and water demands over the past two decades have made urban water supplies a much higher priority than in an earlier era of Colorado River development. Sharp population growth in nearly every urban area served by Colorado River water has caused municipal water managers to think broadly and creatively about efficient water management and ways to limit per capita water uses. States and municipalities across the region have sponsored many creative and useful water conservation, landscaping, and public education programs, but they have not been documented or coordinated in a systematic fashion across the basin or region (the 2003 study from the Western Resource Advocates has been mentioned). Therefore, it is not easy to obtain comprehensive knowledge of the full spectrum of these practices (and successes or setbacks); this state of affairs limits the ability of water managers to learn about experiences from other urban areas in the region. The science and practice in the related fields of demographic projections and regional water demand forecasting lags behind advances in hydrologic and climate sciences. Moreover, linkages between possible changes in climate and Colorado River water availability, and urban water use programs and strategies, have not been comprehensively documented and explored.

Urban water supply-and-demand issues have moved to the fore on the western water landscape. It is therefore important that municipalities and water utilities in the region have good information on future, regional water demands as well as information on successful water utility practices—including drought management—across the region. A thorough investigation of these topics could be used as an action plan to guide and coordinate future urban water management and conservation initiatives across the region.

A comprehensive, action-oriented study of Colorado River region urban water practices and changing patterns of demand should be conducted, as such a study could provide a more systematic basis for water resources planning across the region. At a minimum, the study should address and analyze the following issues:

- historical adjustments to droughts and water shortages,
- demographic projections,
- local and regional water demand forecasting,
- experiences in drought and contingency planning,
- impacts of increasing urban demands on riparian ecology,
- long-term impacts associated with agriculture-urban transfers, and
- contemporary urban water polices and practices (e.g., conservation, landscaping, water use efficiency technologies).

The study could be conducted by the Colorado River basin states, a U.S. federal agency or agencies, a group of universities from across the region, or some combination thereof. The basin states and the U.S. Congress should collaborate on a strategy for commissioning and funding this study. These groups should be prepared to take action based on this study's findings in order to improve the region's preparedness for future inevitable droughts and water shortages.

6

Epilogue

In a series of lectures delivered at the University of Michigan in the late 1980s, Wallace Stegner asked, "What do you do about aridity, if you are a nation inured to plenty and impatient of restrictions and led westward by pillars of fire and cloud? You may deny it for a while. Then you must either adapt to it or try to engineer it out of existence" (Stegner, 1987).

Debates regarding the West's aridity and its implications for urban settlement, population growth, and irrigated agriculture are not new. In the 19th century, John Wesley Powell, Clarence King, and other scientists recognized the unique challenges posed by the region's aridity as they conducted the initial geological surveys of the region. As head of the federal Irrigation Survey in the late 1880s, Powell unsuccessfully challenged Nevada Senator Bill Stewart and influential western ranching and landholding interests when he sought to constrain Western settlement so that land and water resources could first be surveyed and more closely assessed in terms of their carrying capacity. The arid West has also seen ardent advocates of growth and development, such as William Gilpin, first governor of the Colorado territory. Irrigation promoters sharing Gilpin's perspective envisioned a boundless utopia in the western United States, and in fact wished to celebrate "The Blessing of Aridity" (Smythe, 1899). In the 20th century, many writers and historians, including Stegner, Walter Prescott Webb, and Bernard DeVoto reflected eloquently on the region's aridity and on the ways in which settlement patterns were shaped by limited water resources.

For roughly 150 years, the West's aridity has driven both practical development and scientific interest in the Colorado and the region's other major rivers. The limits and values of these watercourses have resulted in many legal agreements and battles over control of their flows; in this regard, the Colorado likely surpasses all other western rivers. With rapid increases in population and water demand in the Colorado River region in the past 25 years, issues of water management, science, and law have assumed tremendous importance and prominence.

The tree-ring-based Colorado River flow reconstructions issued in the late 1990s and early 2000s represent a key advance in scientific understanding of the region's climate and hydrology. General findings from these reconstructions—that sustained, severe droughts have recurred for centuries across the region, that the 1890-1920 period was exceptionally wet, and that the long-term mean flow of the Colorado River is lower than the 15 million acre-feet per year reflected in the Lees Ferry gaged record (and less than the framers of the Colorado River Compact assumed)—are important in themselves. They are also important because their publication coincided with severe drought conditions in the late 1990s and early 2000s, and during a period of increasing population growth and water demands. The convergence of these findings and trends induced great concern among water managers in federal, state, regional, and municipal agencies across the Southwest. Recent water conservation and management initiatives, such as the Department of the Interior's Water 2025 program, acknowledge the challenges and conflicts that will inevitably attend increasing water demands, limited (and likely decreasing) water supplies, and recurrent drought.

Urban water use has always been part of the context of the Law of the River and the operations of Lakes Powell and Mead, but for many decades it was overshadowed by large-scale irrigation development. The 1968 National Research Council report on water and choice in the Colorado River basin, for instance, noted that while population was growing rapidly in the region, "Much of the Colorado basin is almost uninhabited." Large portions of the basin's interior and arid regions remain sparsely populated today, but over the past 40 years, and especially since the mid-1980s, urban water demands within the basin and in water delivery areas outside of the basin have grown in importance in the context of Colorado River water storage and operational decisions. In earlier times, concerns regarding hydroclimatic variability and the Colorado River largely centered on the Bureau of Reclamation's Annual Operating Plan and operational specifics derived from the Law of the River, such as equalization of storage levels between Lakes Powell and Mead. With today's rapidly

Epilogue

growing urban water demands, allocations and obligations defined by the Law of the River are increasingly affected by municipal and industrial water needs. Not only are increasing urban water demands having noticeable effects on reservoir storage levels and instream flows, but the likelihood of further agriculture-urban water transfers promises to shift even more water away from streams and groundwater resources in rural areas.

A future of increasing population growth and urban water demands in a hydroclimatic setting of limited—and likely decreasing—water supplies presents a sobering prospect for elected officials and water managers. If the region's water resources are to be managed sustainably and continue to provide a broad range of benefits to an increasing number of users, the realities of Colorado River water demand and supply will have to be addressed openly and candidly. If the region is to adjust successfully to its rapidly changing water supply-and-demand dynamics, elected and appointed officials, water managers, and the citizenry will require good information on urban water efficiency programs and on options and programs for adjusting to drought. There also needs to be a better appreciation of the economic, social, and environmental impacts that accompany agriculture-urban water transfers.

In the face of these realities and challenges, it is important to acknowledge that the Law of the River, the Colorado River water storage and conveyance infrastructure, and stakeholders tied to the region's economy (including government agencies, farmers, urban water managers, and citizens) have all demonstrated a capacity to cope with water shortages. This report highlights the many factors that are likely to heighten future water management challenges, and which may eventually prompt substantial changes in policies for managing and using water. There is no technical cure-all or panacea capable of resolving the region's increasing water supply-and-demand tensions. As this report notes, future events may necessitate a new level of federal and interstate collaboration on Colorado River water management. Such collaboration may also necessitate more extensive involvement of scientists and engineers with knowledge of water availability and demand trends in formulating water management decisions. The challenges of managing limited water supplies in a region with growing population and demands are not unique to the Colorado River basin, and Colorado River water managers are encouraged to further explore potential benefits that might accrue from scientific exchanges with other regions of the nation and the world.

This report points to several important scientific findings as they relate to Colorado River hydrology and climate. It also includes findings related to cooperation among the basin states and between scientists and water managers. The report recommends that a comprehensive assessment of contemporary urban water management practices and other relevant water supply-and-demand issues be conducted, and that this assessment consider both the full implications of agriculture-to-urban water transfers and future development of regional water demand forecasting. In doing so, it defines an action-oriented study that could provide a more systematic blueprint for improving water management across the rapidly growing and arid Colorado River basin. As the Colorado River basin enters a new phase of coping with aridity and drought, future challenges promise to be more exacting than those faced in the past, and the cooperation that such a study would entail will be of great value. In the 21st century, good scientific information regarding Colorado River flows and variability, and close cooperation and communication at all levels, will prove more important than ever.

References

AAAS (American Association for the Advancement of Science). 2006. AAAS Report XXXI: Research and Development FY 2007. Washington, D.C.: AAAS.

AMTA (American Membrane Technology Association). 2005. How Much Does Desalted Water Cost? Available online at *http://www.membranes-amta.org/media/pdf/desaltingcost.pdf*. Accessed May 16, 2006.

Anderson, L. 2006. The basin states' interim operating alternative. Policy Perspectives 2(3). Available online at *http://www.imakenews.com/cppa/e_article000537691.cfm?x= b11,0,w*. Accessed June 21, 2006.

AWWA (American Water Works Association) and UCAR (University Corporation for Atmospheric Research). 2006. Climate Change and Water Resources: A Primer for Municipal Water Providers. Denver, CO: AWWA.

Billington, D.P., and D.C. Jackson. 2006. Big Dams of the New Deal Era: A Confluence of Engineering and Politics. Norman, OK: University of Oklahoma Press.

Bissell, C.A. 1939. History and First Annual Report of the Metropolitan Water District of Southern California. Los Angeles, CA: Metropolitan Water District.

Brigham, J. 1998. Empowering the West: Electrical Politics before FDR. Lawrence, KS: University Press of Kansas.

Brockway, C.G., and A.A. Bradley. 1995. Errors in streamflow drought statistics reconstructed from tree-ring data. Water Resources Research 31: 2279-2293.

Brooks, D. 1974. Prehistoric soil and water control in the American Southwest: A case study. M.A. Thesis. Flagstaff, AZ: Northern Arizona University.

California Coastal Commission and State Lands Commission. 1999. California Offshore Oil & Gas Leasing and Development Status

Report. Available online at *http://www.coastal.ca.gov/energy/ocs99.pdf.* Accessed July 26, 2006.

California DWR (Department of Water Resources). 2006. Progress on Incorporating Climate Change into Planning and Management of California's Water Resources. Sacramento, CA: California DWR.

California Environmental Protection Agency. 2006. Climate Action Team Report to Governor Schwarzenegger and the Legislature. Available online at *http://www.climatechange.ca.gov/climate_action_team/reports/index.html.* Accessed November 30, 2006.

Cayan, D.R., K.T. Redmond, and L.G. Riddle. 1999. ENSO and hydrologic extremes in the western United States. Journal of Climate 12(9): 2881-2893.

Cayan, D.R., S.A. Kammerdiener, M.D. Dettinger, J.M. Caprio, and D.H. Peterson. 2001. Changes in the onset of spring in the western United States. Bulletin of the American Meteorological Society 82: 399-415.

Christensen, N., and D.P. Lettenmaier. 2006. A multimodel ensemble approach to assessment of climate change impacts on the hydrology and water resources of the Colorado River basin. Hydrology and Earth System Sciences Discussions 3: 3727-3770.

Christensen, N.S., A.W. Wood, N. Voisin, D.P. Lettenmaier, and R.N. Palmer. 2004. The effects of climate change on the hydrology and water resources of the Colorado River basin. Climatic Change 62(1): 337-363.

Colby, B.G., and K.L. Jacobs, eds. 2006. Arizona Water Policy: Management Innovations in an Urbanizing, Arid Region. Washington, D.C.: RFF Press.

Colorado DNR (Department of Natural Resources). 2004. 10-Year Strategic Plan on the Comprehensive Removal of Tamarisk and the Coordinated Restoration of Colorado's Native Riparian Ecosystems. Available online at *http://dnr.state.co.us/NR/rdonlyres/D30D7294-5E0B-4BCB-BE7D-DB2E7F1A1CEE/0/Complete_Tamarisk_10_year_ plan.pdf.* Accessed August 3, 2006.

Cook, E.R., and L.A. Kairiukstis. 1990. Methods of Dendrochronology: Applications in the Environmental Sciences. Dordrecht, The Netherlands: Kluwer Academic Publishers.

Cook, E.R., K.R. Briffa, D.M. Meko, D.S. Graybill, and G. Funkhouser. 1995. The "segment length curse" in long tree-ring chronology development for paleoclimatic studies. The Holocene 5(2): 229-237.

Cooley, H., P.H. Gleick, and G. Wolff. 2006. Desalination with a Grain of Salt: A California Perspective. Oakland, CA: Pacific Institute for Studies in Development, Environment, and Security.

CRS (Congressional Research Service). 1980. State and National Water Use Trends to the Year 2000. Report 96-12. Washington, D.C.: CRS.

Cubash, U., G.A. Meehl, and G.J. Boer. 2001. Projections of future climate change. In Climate Change 2001: The Scientific Basis. Contribution of Working Group I to the Third Assessment Report of the Intergovernmental Panel on Climate Change, J.T. Houghton, Y. Ding, D.J. Griggs, M. Noguer, P.J. van der Linden, X. Dai, K. Maskell, and C.A. Johnson, eds. New York: Cambridge University Press.

CWCB (Colorado Water Conservation Board). 2004. Statewide Water Supply Initiative Report Overview. Available online at *http://cwcb.state.co.us/SWSI/pdfDocs/Report/SWSI8-pgReport Summary3.pdf.* Accessed August 16, 2006.

Dart, A. 1989. Prehistoric Irrigation in Arizona: A Context for Canals and Other Related Cultural Resources. Center for Desert Archaeology Technical Report 89-7. Phoenix, AZ: Arizona State Historic Preservation Office, Arizona State Parks Board.

Davenport, J.H. 2005. Interstate water banking: Evolving Colorado River system agreement. The Water Report 17.

Davis, J. 2006. Arizona's Adequate Water Supply Program: Is it adequate for rural areas? Southwest Hydrology 5(5): 20.

De Stanley, M. 1966. The Salton Sea: Yesterday and Today. Los Angeles, CA: Triumph Press.

Delworth, T.L., and M.E. Mann. 2000. Observed and simulated multidecadal variability in the Northern Hemisphere. Climate Dynamics 16(9): 661-675.

Dettinger, M.D. 2005. From climate-change spaghetti to climate-change distributions for 21st century California. San Francisco Estuary Watershed Science 3(1): 1-14.

DOI (U.S. Department of the Interior). 2003. Water 2025: Preventing Crises and Conflict in the West Interior Proposal Would Concentrate Federal Resources to Support Community Solutions. Available online at *http://www.doi.gov/water2025/news5-2-03.html*. Accessed August 16, 2006.

DOI and USBR (U.S. Bureau of Reclamation). 2006. Colorado River Accounting and Water Use Report: Arizona, California, and Nevada: Calendar Year 2004. Available online at *http://www.usbr.gov/LC/region/g4000/4200Rpts/DecreeRpt/2004/2004.pdf*. Accessed November 16, 2006.

Ebbesmeyer, C.C., D.R. Cayan, D.R. McClain, F.H. Nichols, D.H. Peterson, and K.T. Redmond. 1991. 1976 step in Pacific climate: Forty environmental changes between 1968-1975 and 1977-1984. In Proceedings of the Seventh Annual Pacific Cimate (PACLIM) Workshop, April 1990. California DWR Interagency Ecological Studies Program Technical Report 26, J.L. Betancourt and V.L. Sharp, eds. Sacramento, CA: California Department of Water Resources.

English, M., and S.N. Raja. 1996. Perpective on deficit irrigation. Agricultural Water Management 32: 1-14.

English, M., K.H. Solomon, and G.J. Hoffman. 2002. A paradigm shift in irrigation management. Journal of Irrigation and Drainage 128: 267-277.

Ferrari, R.L. 1988. 1986 Lake Powell Survey: Technical Report REC-ERC-88-6. Denver, CO: Bureau of Reclamation.

Fish, S.K., and P.R. Fish. 1994. Prehistoric desert farmers of the Southwest. Annual Review of Anthropology 23: 83-109.

Fisk, G.G., N.R. Duet, E.H. McGuire, C.E. Angeroth, N.K. Castillo, and C.F. Smith. 2004. Water Resources Data: Arizona, Water Year 2004. Water-Data Report USGS-WRD-AZ-04-1. Available online at *http://pubs.usgs.gov/wdr/2004/wdr-az-04-1/pdf/wdr-az-04-1.pdf*. Accessed August 1, 2006.

Folland, C.K., T.R. Karl, J.R. Christy, R.A. Clarke, G.V. Gruza, J. Jouzel, M.E. Mann, J. Oerlemans, M.J. Salinger, S.W. Wang, et al. 2001. Observed climate variability and change. In Climate Change 2001: The Scientific Basis: Contribution of Working Group I to the Third Assessment Report of the Intergovernmental Panel on Climate Change, J.T. Houghton, Y. Ding, D.J. Griggs, M. Noguer, P.J. van der Linden, X. Dai, K. Maskell, and C.A. Johnson, eds. New York: Cambridge University Press.

Fradkin, P.L. 1984. A River No More: The Colorado River and the West. Tucson, AZ: The University of Arizona Press.

Frederick, K.D. 2001. Water marketing: Obstacles and opportunities. Forum for Applied Research and Public Policy Spring: 54-62.

Fritts, H.C. 1976. Tree Rings and Climate. London: Academic Press.

Fulp, T. 2005a. How low can it go? Southwest Hydrology 4(2): 16-17, 28.

Fulp, T. 2005b. Response of the System to Various Hydrological and Operational Assumptions: Reclamation Modeling Results. Natural Resources Law Center at the University of Colorado at Boulder Twenty-sixth Annual Conference: June 8-10. Available online at *http://wwa.colorado.edu/resources/colorado_river/ hard_times_conference/Fulp_NRLCpresentation.pdf.* Accessed June 22, 2006.

Garrick, D., and K. Jacobs. 2006. Water management on the Colorado River: From surplus to shortage in five years. Southwest Hydrology 5(3): 8-9.

Garstang, M., R. Bruintjes, R. Serafin, H. Orville, B. Boe, W. Cotton, and J. Warburton. 2004. Weather modification: Finding common ground. Bulletin of the American Meteorological Society 86(5): 647-655.

Gavrell, R.C. 2005. Note and Comment: The elephant under the border: An argument for a new, comprehensive treaty for the transboundary waters and aquifers of the United States and Mexico. Colorado Journal of International Environmental Law and Policy 16(1): 189, 192-193.

Glantz, M. 1988. Societal Responses to Regional Climate Change: Forecasting by Analogy. Boulder, CO: Westview Press.

Gleick, P., ed. 2000. Water: The Potential Consequences of Climate Variability and Change for the Water Resources of the United States. The Report of the Water Sector Assessment Team of the National Assessment of the Potential Consequences of Climate Variability and Change for the U.S. Global Change Research Program. Available online at *http://www.usgcrp.gov/usgcrp/nacc/ water/default.htm.* Accessed November 30, 2006.

Gleick, P., H. Cooley, and D. Groves. 2005. California Water 2030: An Efficient Future. Oakland, CA: Pacific Institute for Studies in Development, Environment, and Security.

Gloss, S.P., J.E. Lovich, and T.S. Melis. 2005. The State of the Colorado River Ecosystem in Grand Canyon: USGS Circular 1282. Reston, VA: U.S. Geological Survey.

Gottlieb, R., and M. Fitzsimmons. 1991. Thirst for Growth: Water Agencies as Hidden Government in California. Tucson, AZ: University of Arizona Press.

Governors. 2005. Letter to Secretary of the Interior Gale A. Norton from the States of Arizona, California, Colorado, Nevada, New Mexico, Utah, and Wyoming Governor's Representatives on Colorado River Operations.

Graumlich, L.G., and L.B. Brubaker. 1986. Reconstruction of annual temperature (1590-1979) for Longmire, Washington, derived from tree rings. Quaternary Research 25: 223-234.

Gray, S.T., L.J. Graumlich, J.L. Betancourt, and G.D. Pederson. 2004. A tree-ring based reconstruction of the Atlantic Multidecadal Oscillation since 1567 A.D. Geophysical Research Letters 31: L12205, doi: 10.1029/2004GL019932.

Griles, J.S. 2004. Building on Success, Facing the Challenges Ahead. Speech to the Colorado River Water Users Association. Available online at *http://www.doi.gov/news/041217speech*. Accessed July 20, 2006.

Hamlet, A.F., and D.P. Lettenmaier. 1999a. Effects of climate change on hydrology and water resources in the Columbia River basin. Journal of the American Water Resources Association 35: 1597-1623.

Harding, B. 2006. Presentation: Counting Cards in Hydrology: Can We Beat the House as We Bet on the Future? Boulder, CO: Hydrosphere Resource Consultants.

Heim, R.R., Jr. 2002. A review of twentieth century drought indices used in the United States. Bulletin of the American Meteorological Society 83(8): 1149-1165.

Hely, A.J. 1969. Lower Colorado River water supply: Its magnitude and distribution. U.S. Geological Survey Professional Paper 486-D. Washington, D.C.: USGS.

Hidalgo, H.G. 2004. Climate precursors of multidecadal drought variability in the western United States. Water Resources Research 40: W12504, doi:10.1029/2004WR003350.

Hidalgo, H.G., T.C. Piechota, and J.A. Dracup. 2000. Alternative principal components regression procedures for dendrohydrologic reconstructions. Water Resources Research 36(11): 3241-3249.

Hirschboeck, K.K., and D.M. Meko. 2005. A tree-ring based assessment of synchronous episodes in the upper Colorado and Salt-Verde-Tonto river basins. Available online at *http://fp.arizona.*

edu/khirschboeck/SRP%20WEB/Final.Report/Final.Final.Report. pdf. Accessed May 16, 2005.

Hoerling, M.P., and A. Kumar. 2003. The perfect ocean for drought. Science 299(5607): 691-694.

Holling, C.S., ed. 1978. Adaptive Environmental Assessment and Management. New York: John Wiley & Sons and International Institute for Applied Systems Analysis.

Holt, J.K., H.G. Park, Y. Wang, M. Staderman, A.B. Artyukhin, C.P. Grigoropoulos, A. Noy, and O. Jakajin. 2006. Fast mass transport through sub-2-nanometer carbon nanotubes. Science 312: 1034-1037.

Houghton, J. 2004. Global Warming: The Complete Briefing, 3rd edition. New York: Cambridge University Press.

Howe, C.W., J.K. Lazo, and K.R. Weber. 1990. The economic impacts of agriculture-to-urban water transfers on the area of origin: A case study of the Arkansas River Valley in Colorado. American Journal of Agricultural Economics 72(5): 1200-1204.

Hundley, N. 1966. Dividing the Waters: A Century of Conflict Between the United States and Mexico. Berkeley, CA: University of California Press.

Hundley, N. 1975. Water and the West: The Colorado Compact and the Politics of Water in the American West. Berkeley, CA: University of California Press.

Hundley, N. 1986. The West against itself: The Colorado River–an institutional history. In New Courses for the Colorado River: Major Issues for the Next Century, G.D. Weatherford and F.L. Brown, eds. Albuquerque, NM: University of New Mexico Press.

Hundley, N. 1992. The Great Thirst: Californians and Water, 1770s-1990s. Berkeley, CA: University of California Press.

IBWC (International Boundary and Water Commission). 1973. Minute 242: Permanent and Definitive Solution to the International Problem of the Salinity of the Colorado River. Available online at *http://www.ibwc.state.gov/Files/Minutes/Min242.pdf.* Accessed May 15, 2006.

Ingram, H., A.D. Tarlock, and C.R. Oggins. 1991. The law and politics of the operation of Glen Canyon Dam. In Colorado River Ecology and Dam Management. Proceedings of a Symposium May 24-25, 1990, New Mexico. Washington, D.C.: National Academy Press.

Jenkins, M. 2005. What's worse than the worst-case scenario? Real life. High Country News 37(5): 13.

Jurriens, M., P.P. Mollinga, and P. Wester. 1996. Scarcity by Design: Protective Irrigation in India and Pakistan. Wageningen, The Netherlands: International Institute for Land Reclamation and Improvement.

Kahrl, W.L. 1982. Water and Power: The Conflict over Los Angeles Water Supply in the Owens Valley. Berkeley, CA: University of California Press.

Kelly, S., and H. Ayers. 1982. Water Conservation Alternatives for California: A Micro-economic Analysis. ERS Staff Report AGES820417. Washington, D.C.: U.S. Department of Agriculture.

Kenney, D.S., R.A. Klein, and M.P. Clark. 2004. Use and effectiveness of municipal water restrictions during drought in Colorado. Journal of the American Water Resources Association 40(1): 77-87.

Keyantash, J., and J.A. Dracup. 2002. The quantification of drought: An evaluation of drought indices. Bulletin of the American Meteorological Society 83(8): 1167-1180.

Keys, J. 2006. Cooperation key to Reclamation's future water works. Southwest Hydrology 5(2): 26-27.

Kittel, T.G.F., J.A. Royle, C. Daly, N.A. Rosenbloom, S.P. Gibson, H.H. Fisher, D.S. Schimel, L.M. Berliner, and VEMAP Participants. 1997. A gridded historical (1895-1993) bioclimate dataset for the conterminous United States. 10th AMS Conference on Applied Climatology, Reno, NV, October 20-23, 1997.

Kleinsorge, P.L. 1941. Boulder Canyon Project: Historical and Economic Aspects. Palo Alto, CA: Stanford University Press.

Knight, J.R., R.J. Allan, C.K. Folland, M. Vellinga, and M.E. Mann. 2005. A signature of persistent natural thermohaline circulation cycles in observed climate. Geophysical Research Letters 32: L20708, doi:10.1029/2005GL024233.

Krishnamurti, T.N., C.M. Kishtawal, T.E. LaRow, D.R. Bachiochi, Z. Zhang, C.E. Willifor, S. Gadgil, and S. Surendran. 1999. Improved weather and seasonal climate forecasts from multimodel superensemble. Science 285: 1548-1550.

LaRue, E.C. 1916. Colorado River and Its Utilization. USGS Water-Supply Paper 395. Washington, D.C.: Government Printing Office.

References

Lee, K.N. 1999. Appraising Adaptive Management. Conservation Ecology 3(2):3; available online at *http://www.consecol.org/vol3/iss2/art3/*. Accessed October 3, 2006.

Loaiciga, H.A., L. Haston, and J. Michaelsen. 1993. Dendrohydrology and long-term hydrologic phenomena. Reviews of Geophysics 31: 151-171.

Lord, W.B., J.F. Booker, D.M. Getches, B.L. Harding, D.S. Kenney, and R.A. Young. 1995. Managing the Colorado River in a severe sustained drought: An evaluation of institutional options. Water Resources Bulletin 31(5): 939-944.

MacDonnell, L., D. Getches, and W. Hugenberg, Jr. 1995. The law of the Colorado River: Coping with severe sustained drought. Water Resources Bulletin 31(5): 825-836.

Manguerra, H.B., and L.A. Garcia. 1996. Drainage and no-drainage cycles for salinity management in irrigated areas. Transactions of the ASAE 39(6): 2039-2049.

Mantua, N.J., S.R. Hare, Y. Zhang, J.M. Wallace, and R.C. Francis. 1997. A Pacific interdecadal climate oscillation with impacts on salmon production. Bulletin of the American Meteorological Society 78: 1069-1079.

Marston, E. 1987. Reworking the Colorado River basin. In Western Water Made Simple. Washington, D.C.: Island Press.

Martin, R. 1989. The Story that Stands Like a Dam: Glen Canyon and the Struggle for the Soul of the West. New York: Henry Holt.

Maurer, E.P. 2007. Uncertainty in hydrologic impacts of climate change in the Sierra Nevada, California under two emissions scenarios. Climatic Change, doi:10.1007/s10584-006-9180-9; available online at *http://www.engr.scu.edu/~emaurer/papers/maurer_uncertainties_ca_cc_2007.pdf*. Accessed March 27, 2007.

Maurer, E.P., A.W. Wood, J.C. Adam, D.P. Lettenmaier, and B. Nijssen. 2002. A long-term hydrologically-based data set of land surface fluxes and states for the conterminous United States. Journal of Climate 15: 3237-3251.

McCabe, G.J., and M.D. Dettinger. 1999. Decadal variations in the strength of ENSO teleconnections with precipitation in the western United States. International Journal of Climatology 19(13): 1399-1410.

McCabe, G.J. and M.A. Palecki. 2006. Multidecadal climate variability of global lands and oceans. International Journal of Climatology 26(7): 849.

McCabe, G.J., M.A. Palecki, and J.L. Betancourt. 2004. Pacific and Atlantic Ocean influences on multidecadal drought frequency in the United States. Proceedings of the National Academy of Sciences 101: 4136-4141.

Medina, J.G., and J. Gumper. 2004. YARDX: Yield and Reliability Demonstrated in Xeriscape: Final Report. Littleton, CO: Metro Water Conservation.

Meko, D.M. 2005. Changes in regional hydroclimatology and water resources on seasonal to interannual and decade-to-century timescales. In Encyclopedia of Hydrological Sciences, Volume 5, Part 17, Climate Change, M.G. Anderson, ed. Hoboken, NJ: John Wiley & Sons.

Meko, D.M., and D.A. Graybill. 1995. Tree-ring reconstruction of Upper Gila River discharge. Water Resources Bulletin 31: 605-616.

Meko, D.M., C.W. Stockton, and W.R. Boggess. 1995. The tree-ring record of severe sustained drought. Water Resources Bulletin 31: 789-801.

Meko, D.M., M.D. Therrell, C.H. Baisan, and M.K. Hughes. 2001. Sacramento River flow reconstructed to AD 869 from tree rings. Journal of the American Water Resources Association 37: 1029-1039.

Meyer, M.C. 1984. Water in the Hispanic Southwest: A Social and Legal History, 1550-1850. Tuscon, AZ: The University of Arizona Press.

Meyers, C.J. 1967. The Colorado basin. In The Law of International Drainage Basins, A.H. Garretson, R.D. Hayton, and C.J. Olmstead, eds. Dobbs Ferry, NY: Oceana Publications.

Michaelsen, J. 1987. Cross-validation in statistical climate forecast models. Journal of Climate and Applied Meteorology 26: 1589-1600.

Michaelsen, J., H.A. Loaiciga, L. Haston, and S. Garver. 1990. Estimating Drought Probabilities in California Using Tree Rings. California Department of Water Resources Report B-57105. Santa Barbara, CA: University of California.

Milly, P.C.D., K.A. Dunne, and A.V. Vecchia. 2005. Global pattern of trends in streamflow and water availability in a changing climate. Nature 438: 347-350.

Moeller, B. 1971. Phil Swing and Boulder Dam. Berkeley, CA: University of California Press.

Mote, P.W., A.F. Hamlet, M.P. Clark, and D.P. Lettenmeier. 2005. Declining mountain snowpack in western North America. Bulletin of the American Meteorological Society 86(1): 39-49.

Murphy, D.F. 2003. Pact in West Will Send Farms' Water to Cities. New York Times, October 17, p. A1.

Nash, L. 1991. The implications of climatic change for streamflow and water supply in the Colorado River basin. In Managing Water Resources in the West under Conditions of Climate Uncertainty: A Proceedings. Washington, D.C.: National Academy Press.

Nash, L.L., and P. Gleick. 1991. The sensitivity of streamflow in the Colorado basin to climatic changes. Journal of Hydrology 125: 221-241.

Nash, L.L., and P. Gleick. 1993. The Colorado River Basin and Climate Change: The Sensitivity of Streamflow and Water Supply to Variations in Temperature and Precipitation. Report EPA 230-R-93-009. Washington, D.C.: Environmental Protection Agency.

Nash, R. 1967. Wilderness and the American Mind. New Haven, CT: Yale University Press.

NPS (National Park Service). 2005. Fight the Invasion: Controlling Invasive Plant Species at Grand Canyon National Park. Grand Canyon, AZ: NPS, Grand Canyon National Park Foundation, and The Arizona Water Protection Fund Commission.

NPS. 2006. Tamarisk Management and Tributary Restoration. Available online at *http://www.nps.gov/archive/grca/compliance/pdf/tam-site.pdf*. Accessed September 6, 2006.

NRC (National Research Council). 1964. Scientific Problems of Weather Modification. Washington, D.C.: National Academy of Sciences.

NRC. 1966. Weather and Climate Modification, Problems and Prospects. Washington, D.C.: National Academy of Sciences.

NRC. 1968. Water and Choice in the Colorado Basin: An Example of Alternatives in Water Management. Washington, D.C.: National Academy of Sciences.

NRC. 1973. Weather and Climate Modification. Washington, D.C.: National Academy of Sciences.

NRC. 1987. River and Dam Management: A Review of the Bureau of Reclamation's Glen Canyon Environmental Studies. Washington, D.C.: National Academy Press.

NRC. 1991a. Colorado River Ecology and Dam Management. Proceedings of a Symposium, May 24-25, 1990, Santa Fe, NM. Washington, D.C.: National Academy Press.

NRC. 1991b. Managing Water Resources in the West under Conditions of Climate Uncertainty: A Proceedings. Washington, D.C.: National Academy Press.

NRC. 1992. Water Transfers in the West: Efficiency, Equity, and the Environment. Washington, D.C.: National Academy Press.

NRC. 1996. River Resource Management in the Grand Canyon. Washington, D.C.: National Academy Press.

NRC. 1998. Issues in Potable Water Reuse: The Viability of Augmenting Drinking Water Supplies with Reclaimed Water. Washington, D.C.: National Academy Press.

NRC. 1999. Downstream: Adaptive Management of Glen Canyon Dam and the Colorado River Ecosystem. Washington, D.C.: National Academy Press.

NRC. 2003. Critical Issues in Weather Modification Research. Washington, D.C.: National Academy Press.

NRC. 2004. Review of the Desalination and Water Purification Technology Roadmap. Washington, D.C.: National Academies Press.

NRC. 2006. Surface Temperature Reconstructions for the Last 2,000 Years. Washington, D.C.: National Academies Press.

NSB (National Science Board). 1966. Weather and Climate Modification. NSF-66-3. Arlington, VA: National Science Foundation.

Orville, H.D., B.A. Boe, G.W. Bomar, W.R. Cotton, B.L. Marler, and J.A. Warburton. 2004. A Response by the Weather Modification Association to the National Research Council's Report Titled "Critical Issues in Weather Modification Research." Available online at *http://www.weathermodification.org/images/FinalReport.pdf*. Accessed May 15, 2006.

Pankratz, T., and J. Tonner. 2003. Desalination.com: An Environmental Primer. Houston, TX: Lone Oak Publishing.

Payne, J.T., A.W. Wood, A.F. Hamlet, R.N. Palmer, and D.P. Lettenmaier. 2004. Mitigating the effects of climate change on the water resources of the Columbia River basin. Climatic Change 62: 233-256.

References

Pearson, B.E. 2002. Still the Wild River Runs: Congress, the Sierra Club, and the Fight to Save Grand Canyon. Tucson, AZ: University of Arizona Press.

Piechota, T.C., J. Timilsena, G. Tootle, and H. Hidlago. 2004. The western U.S. drought: How bad is it? EOS 85(32): 301-308.

Powell, J.W. 1895. Canyons of the Colorado. New York: Flood & Vincent.

Raley, B.F. 2001. Private irrigation and the Grand Valley Irrigation Company. In Fluid Arguments: Five Centuries of Western Water Conflict, C. Miller, ed. Tucson, AZ: University of Arizona Press.

Redmond, K.T. 2002. The depiction of drought: A commentary. Bulletin of the American Meteorological Society 83(8): 1143-1147.

Redmond, K.T. In press. Climate variability and change as a backdrop for western resource management. In USDA Forest Service Pacific Northwest Research Station Technical Report, Bringing Climate into Natural Resource Management. Proceedings of a USDA Conference, May 28-30, 2005, Portland, OR.

Redmond, K.T., and R.W. Koch. 1991. Surface climate and streamflow variability in the western United States and their relationship to large-scale circulation indexes. Water Resources Research 27: 2381-2399.

Regonda, S.K., B. Rajagopalan, M. Clark, and J. Pitlick. 2005. Seasonal cycle shifts on hydroclimatology over the western United States. Journal of Climate 18: 372-384.

Reisner, M. 1986. Cadillac Desert: The American West and Its Disappearing Water. New York: Viking Penguin.

Revelle, R.R., and P.E. Waggoner. 1983. Effects of carbon dioxide-induced climate change on water supplies in the western United States. In Changing Climate. Washington, D.C.: National Academy Press.

Rodebaugh, D. 2005. Red-letter day for A-LP: Fulfilling Indian water claims still years away. The Durango Herald, August 13.

Rogers, P. 1993. America's Water: Federal Roles and Responsibilities. Cambridge, MA: MIT Press.

Schubert, S.D., M.J. Suarez, P.J. Pegion, R.D. Koster, and J.T. Bacmeister. 2004a. Causes of long-term drought in the U.S. Great Plains. Journal of Climate 17(3): 485-503.

Schubert, S.D., M.J. Suarez, P.J. Pegion, R.D. Koster, and J.T. Bacmeister. 2004b. On the cause of the 1930s Dust Bowl. Science 303: 1855-1859.

Schulman, E. 1956. Dendroclimatic Changes in Semiarid America. Tucson, AZ: University of Arizona Press.

Seager, R. 2006. Persistent Drought in North America: A Climate Modeling and Paleoclimate Perspective. Available online at *http://www.ldeo.Columbia.edu/res/div/ocp/drought.* Accessed October 6, 2006.

Seager, R., Y. Kushnir, C. Herweijer, N. Naik, and J. Velez. 2005. Modeling of tropical forcing of persistent droughts and pluvials over western North America: 1856-2000. Journal of Climate 18: 4065-4088.

Smith, L.P., and C.W. Stockton. 1981. Reconstructed streamflow for the Salt and Verde rivers from tree-ring data. Water Resources Bulletin 17(6): 939-947.

Smythe, W.E. 1899. The Conquest of Arid America. New York: Harper and Brothers.

SNWA (Southern Nevada Water Authority). 2004. Five-Year Conservation Plan 2004-2009. Available online at *http://www.snwa.com/assets/pdf/consv_plan_intro.pdf.* Accessed May 24, 2006.

SNWA. 2006. Chapter 3: The SNWA water resource portfolio. In Southern Nevada Water Authority 2006 Water Resource Plan. Available online at *http://www.snwa.com/assets/pdf/wr_plan06_chapter3.pdf.* Accessed December 6, 2006.

Starr, K. 1990. Material Dreams: Southern California Through the 1920s. New York: Oxford University Press.

Stegner, W. 1954. Beyond the Hundredth Meridian: John Wesley Powell and the Second Opening of the West. Boston, MA: Houghton Mifflin.

Stegner. W. 1987. The American West as Living Space. Ann Arbor, MI: University of Michigan Press.

Stevens, J.E. 1988. Hoover Dam: An American Adventure. Norman, OK: University of Oklahoma Press.

Stewart, I.T., D.R. Cayan, and M.D. Dettinger. 2005. Changes toward earlier streamflow timing across western North America. Journal of Climate 18(8): 1136-1155.

Stockton, C.W. 1975. Long-term streamflow records reconstructed from tree rings. Papers of the Laboratory of Tree-Ring Research 5. Tucson, AZ: University of Arizona Press.

Stockton, C.W., and W.R. Boggess. 1979. Geohydrological Implications of Climate Change on Water Resource Development. Fort Belvoir, VA: U.S. Army Coastal Engineering Research Center.

Stockton, C.W., and G.C. Jacoby. 1976. Long-term surface-water supply and streamflow trends in the Upper Colorado River basin based on tree-ring analyses. Lake Powell Research Project Bulletin 18: 1-70.

Stokes, M.A., and T.L. Smiley. 1968. An Introduction to Tree-Ring Dating. Tucson, AZ: University of Arizona Press.

Sturgeon, S. 2002. The Politics of Western Water: The Congressional Career of Wayne Aspinall. Tucson, AZ: University of Arizona Press.

Sutton, R.T., and D.L.R. Hodson. 2005. Atlantic Ocean forcing of North American and European summer climate. Science 309: 115-118.

Svoboda, M., D. LeComte, M. Hayes, R. Heim, K. Gleason, J. Angel, B. Rippey, R. Tinker, M. Palecki, D. Stooksbury, D. Miskus, and S. Stephens. 2002. The Drought Monitor. Bulletin of the American Meteorological Society 83(8): 1181-1190.

Tarlock, A.D. 1991. Western water law, global climate change, and risk allocation. In Managing Water Resources in the West under Conditions of Climate Uncertainty: A Proceedings. Washington, D.C.: National Academy Press.

Think New Mexico. 2003. Rio Vivo! The Need for a Strategic River Reserve in New Mexico. Santa Fe, NM: Think New Mexico.

Topping, D.J., J.C. Schmidt, and L.E. Vierra, Jr. 2003. Computation and analysis of the instantaneous-discharge record for the Colorado River at Lees Ferry, Arizona–May 8, 1921, though September 30, 2000. Professional Paper 1677. Reston, VA: USGS.

Trenberth, K.E., and J.W. Hurrell. 1994. Decadal atmospheric-ocean variations in the Pacific. Climate Dynamics 9(6): 303-319.

Trimmer, W.L. 1990. Applying partial irrigation in Pakistan. Journal of Irrigation and Drainage 116(3): 342-353.

UCRSFIG (Upper Colorado Region State-Federal Interagency Group). 1971. Upper Colorado Region Comprehensive Framework Study. Appendix V: Water Resources. Washington, D.C.: U.S. Water Resources Council.

USBR (U.S. Bureau of Reclamation). 1947. Colorado River: A Natural Menace Becomes a Natural Resource. Washington, D.C.: Government Printing Office.

USBR. 1977. Colorado River System Consumptive Uses and Losses Report: 1971-1975. Available online at *http://www.usbr.gov/uc/library/envdocs/reports/crs/crsul.html*. Accessed December 11, 2006.

USBR. 1981. Colorado River System Consumptive Uses and Losses Report: 1976-1980. Available online at *http://www.usbr.gov/uc/library/envdocs/reports/crs/crsul.html*. Accessed December 11, 2006.

USBR. 1991. Colorado River System Consumptive Uses and Losses Report: 1981-1985. Available online at *http://www.usbr.gov/uc/library/envdocs/reports/crs/crsul.html*. Accessed December 11, 2006.

USBR. 1995. Operation of Glen Canyon Dam: Colorado River Storage Project, Arizona: Final Environmental Impact Statement. Available online at *http://www.usbr.gov/uc/library/evdocs/eis/gc/glen.html*. Accessed October 4, 2006.

USBR. 1998. Colorado River System Consumptive Uses and Losses Report: 1986-1990. Available online at *http://www.usbr.gov/uc/library/envdocs/reports/crs/crsul.html*. Accessed December 11, 2006.

USBR. 2002. Colorado River System Consumptive Uses and Losses Report: 1991-1995. Available online at *http://www.usbr.gov/uc/library/envdocs/reports/crs/crsul.html*. Accessed December 11, 2006.

USBR. 2004. Colorado River System Consumptive Uses and Losses Report: 1996-2000. Available online at *http://www.usbr.gov/uc/library/envdocs/reports/crs/crsul.html*. Accessed December 11, 2006.

USGS (U.S. Geological Survey). 1954. Compilation of Records of Surface Waters of the United States through September 1950. USGS Water-Supply Paper 1313. Washington, D.C.: Government Printing Office.

USGS. 2002. Precipitation History of the Colorado Plateau Region, 1900-2000. U.S Geological Survey Fact Sheet 119-02. Available online at *http://pubs.usgs.gov/fs/2002/fs119-02/*. Accessed December 4, 2006.

References

USGS. 2003. Ground-Water Depletion Across the Nation. U.S Geological Survey Fact Sheet 103-03. Available online at *http://pubs.usgs.gov/fs/fs-103-03/*. Accessed November 16, 2006.

USGS. 2005. The State of the Colorado River Ecosystem in Grand Canyon: A Report of the Grand Canyon Monitoring and Research Center 1991-2004. USGS Circular 1282. Reston, VA: USGS.

Walker, G.T. 1925. On periodicity (with discussion). Quarterly Journal of the Royal Meteorological Society 51: 337-346.

Walters, C. 1986. Adaptive Management of Renewable Resources. New York: Macmillan.

Wangnick, K. 2002. IDA Worldwide Desalting Plants Inventory. Report 17. Gnarrenburg, Germany: Wangnick GMBH.

Waters, F. 1946. The Colorado (Rivers of America Series). New York: Rinehart.

WCD (World Commission on Dams). 2000. Dams and Development: A New Framework for Discussion. London: Earthscan Publications.

Webb, R.H., J.C. Schmidt, G.R. Marzolf, and R.A. Valdez, eds. 1999. The Controlled Flood in Grand Canyon. Washington, D.C.: American Geophysical Union.

WGA (Western Governors' Association). 2004. Creating a drought early warning system for the 21st Century: The National Integrated Drought Information System. Western Governors Association, endorsed June 2004. Available online at *http://www.west gov.org/wga/publicat/nidis.pdf*. Accessed July 22, 2006.

WGA. 2006. Water Needs and Strategies for a Sustainable Future. Denver, CO: WGA.

Wichelns, D. 2002. An economic perspective on the potential gains from improvements in irrigation water management. Agricultural Water Management 52: 233-248.

Wigley, T., K. Briffa, and P.D. Jones. 1984. On the average value of correlated time series, with applications in dendroclimatology and hydrometeorology. Journal of Climate and Applied Meteorology 23: 201-213.

Wilhite, D.A. 1997a. Responding to drought: Common threads from the past, visions for the future. Journal of the American Water Resources Association 33(5): 951-959.

Wilhite, D.A. 1997b. Improving drought management in the West: The role of mitigation and preparedness. Report to the Western

Water Policy Review Advisory Commission. Lincoln, NE: University of Nebraska National Drought Mitigation Center.

Wilhite, D.A., and M.H. Glantz. 1985. Understanding the drought phenomenon: The role of definitions. Water International 10(3): 111-120.

WMA (Weather Modification Association). 2005. Weather Modification in the United States. Fresno, CA: WMA.

Woodhouse, C. A., and J.J. Lukas. 2006. Multi-century tree-ring reconstructions of Colorado streamflow for water resource planning. Climatic Change, doi: 10.1007/s10584-006-9055-0.

Woodhouse, C.A., and D.M. Meko. 2007. Dendroclimatology, dendrohydrology, and water resources management. In Dendroclimatology: Progress & Prospects, M.K. Hughes, T.W. Swetman, and H.F. Diaz, eds. New York: Springer.

Woodhouse, C.A., S.T. Gray, and D.M. Meko. 2006. Updated streamflow reconstructions for the Upper Colorado River basin. Water Resources Research 42: W05415, doi: 10.1029/2005WR 004455.

Worster, D. 1985. Rivers of Empire: Water Aridity, and the Growth of the American West. New York: Pantheon Press.

Worster, D. 2000. A River Running West: The Life of John Wesley Powell. New York: Oxford University Press.

WRA (Western Resource Advocates). 2003. Smart Water: A Comparative Study of Urban Water Use Efficiency Across the Southwest. Boulder, CO: Western Resource Advocates.

WWDC (Wyoming Water Development Commission). 2006. Wyoming Weather Modification Five-Year Pilot Program: Medicine Bow and Sierra Madre Mountain Project, Wind River Range Project. Cheyenne, WY: WWDC.

Zarbin, E. 1984. Roosevelt Dam: A History to 1911. Phoenix, AZ: Salt River Project.

Zarbin, E. 1997. Two Sides of the River: Salt River Valley Canals, 1867-1902. Phoenix, AZ: Salt River Project.

Appendixes

Appendix A

**The States of Arizona, California, Colorado, Nevada,
New Mexico, Utah and Wyoming
Governor's Representatives on Colorado River Operations**

February 3, 2006

Honorable Gale A. Norton, Secretary
Department of the Interior
1849 C. Street, NW
Washington, D.C. 20240

Re: Development of Lower Basin Shortage Guidelines and Coordinated Management Strategies for the Operation of Lake Mead and Lake Powell Under Low Reservoir Conditions

Dear Secretary Norton:

The materials attached to this letter contain descriptions of the programs that the seven Colorado River Basin States suggest be included within the scope of the environmental impact statement (EIS) for the proposed *Colorado River Reservoir Operations: Development of Lower Basin Shortage Guidelines and Coordinated Management Strategies for Lake Powell and Lake Mead Under Low Reservoir Conditions* (70 Fed. Reg. 57322) (Sept. 30, 2005).

The Basin States, Bureau of Reclamation and others have consulted regularly since our previous correspondence on August 25, 2005 to further discuss and refine recommended management strategies for the Colorado River system. Subsequently, individual entities within the seven Basin States submitted oral and written comments to the Bureau of Reclamation on the above-referenced EIS process. Attachment A, "Seven Basin States' Preliminary Proposal Regarding Colorado River Interim Operations," is submitted as a consensus document on behalf of the seven Basin States. Please recognize that the States are still actively working on the matters addressed in this submission and anticipate further refinement.

Our recommendation is designed to provide input for the Department's consideration as it develops additional operational and water accounting procedures to: 1) delay the onset and minimize the extent and duration of shortages in the Lower Division States; 2) maximize the protection afforded the Upper Division States by storage in Lake Powell against possible curtailment of Upper Basin uses; 3) provide for more efficient, flexible, responsive and reliable operation of the system reservoirs for the benefit of both the Upper and Lower Basins by developing additional system water supplies through extraordinary conservation, system efficiency and augmentation projects; 4) allow the continued development and use of the Colorado River resource in both the Upper and Lower Basins; and 5) allow for development of dedicated water supplies through participation in improvements to system efficiency and clarification of how to proceed with development of non-system water reaching the Lower Basin

The Honorable Gale A. Norton
February 3, 2006
Page 2 of 3

mainstream. It is our position that implementation of these operational and accounting procedures can be accomplished without modification of the Long Range Operating Criteria or other elements of the law of the river.

The States' attached proposal incorporates an approach to shortage management. Additionally, the proposal includes modification and extension of the Department's Interim Surplus Guidelines to incorporate operations for all reservoir conditions.

The attached proposal also addresses the States' recommended approach to implementation of shortages pursuant to the U.S.-Mexico Treaty of 1944. We request that the Department of the Interior initiate, at the earliest appropriate time, consultation with the U.S. Section of the International Boundary and Water Commission and the U.S. Department of State on implementation of Treaty shortages. We further request the opportunity to consult with Interior and State Department officials on this issue as the federal government formulates its approach to any bi-national consultation with Mexico.

An agreement between Basin State water managers and users will be necessary to put in place additional terms upon which they have reached common understanding. We intend that this agreement be finalized while Reclamation is preparing the draft EIS, and be executed as soon as practicable. We are including with this letter a draft version of the agreement (Attachment B), to memorialize our current understandings and to provide you the benefits of our thoughts at this time. As with Attachment A, please recognize that the parties are still actively working on the matters addressed in Attachment B, and contemplate additional development and refinement of the agreement. We recognize that timely execution of our agreement is necessary in order to allow funding of certain efficiency projects to go forward.

During the time Reclamation is preparing the draft EIS, the States will move forward with a package of other actions that include implementation of a demonstration program for extraordinary conservation in 2006, system efficiency projects, preparation of an action plan for system augmentation through weather modification, execution of a memorandum of understanding for preparing a Lower Division States interstate drought management plan, development of forbearance agreements among the Lower Division States and the initiation of a study for long-term augmentation of Colorado River system water supplies. The States have already begun the consultant procurement process to support the long-term augmentation study, and intend to complete a weather modification action plan and a memorandum of understanding for interstate drought planning as soon as practicable. The Basin States recognize that Reclamation is undertaking NEPA compliance separately to determine whether to construct a regulating reservoir near Drop 2 of the All-American Canal and urge swift completion of that process.

We appreciate the opportunity you have provided for the Colorado River Basin States to recommend to you a program of reservoir management that considers all their respective concerns and interests. The Basin States look forward to working with you and Reclamation in analyzing and addressing these matters.

Appendix A

The Honorable Gale A. Norton
February 3, 2006
Page 3 of 3

Sincerely,

Herb Guenther
Director
Arizona Department of Water Resources

Scott Balcomb
Governor's Representative
State of Colorado

Richard Bunker
Chairman
Colorado River Commission of Nevada

John R. D'Antonio, Jr.
Governor's Representative
State of New Mexico

D. Larry Anderson
Director
Utah Division of Water Resources

Gerald R. Zimmerman
Executive Director
Colorado River Board of California

Rod Kuharich
Director
Colorado Water Conservation Board

Patricia Mulroy
General Manager
Southern Nevada Water Authority

Patrick Tyrrell
State Engineer
State of Wyoming

List of Attachments:

Attachment A: Seven Basin States' Preliminary Proposal Regarding Colorado River Interim Operations

Attachment B: Draft Agreement

Attachment A

Seven Basin States' Preliminary Proposal Regarding Colorado River Interim Operations

The Seven Basin States (States) have worked together to recommend interim operations to the Secretary that should minimize shortages in the Lower Basin and avoid the risk of curtailment in the Upper Basin through conservation, more efficient reservoir operations, and long-term alternatives to bring additional water into the Colorado River community.

The States' recommendation has three key elements. First, the States propose to manage the reservoirs to minimize shortages and avoid curtailments. Second, the States have identified actions in the Lower Basin to conserve water. Third, the States recommend a specific proposal for implementing shortages in the Lower Basin. Finally, the States recognize the need for additional water supplies to meet the current and future needs in the Basin.

Section 1. Allocation of Unused Basic Apportionment Water under Article II(B)(6)

A. Introduction

Article II(B)(6) of the 1964 Decree in *Arizona* v. *California* (Decree) allows the Secretary to allocate water that is apportioned to one Lower Division State, but is for any reason unused in that State, to another Lower Division State. This determination is made for one year only and no rights to recurrent use of the water accrue to the State that receives the allocated water.

B. Application of Unused Basic Apportionment

Before making a determination of a surplus condition under this proposal, the Secretary will determine the quantity of apportioned but unused water under Article II (B)(6), and will allocate such water in the following order of priority.

1. Meet the direct delivery domestic use requirements of the Metropolitan Water District of Southern California, (MWD) and the Southern Nevada Water Authority (SNWA), as allocated between them by agreement.

2. Meet the needs of off stream banking activities by MWD in California and SNWA in Nevada, as allocated between them by agreement.

3. Meet the other needs for water in California in accordance with the California Seven-Party Agreement as supplemented by the Quantification Settlement Agreement.

Section 2. Coordinated Operation of Lakes Powell and Mead

Figure 1 describes the operating strategy that has been agreed to by the Colorado River Basin States.

Appendix A

Powell Elevation (feet)	Powell Operation	Powell Live Storage (maf)
3700	Equalize or 8.23 maf	24.32
3636 - 3664 (see table below)		15.54 - 19.02 (2008 - 2025)
	8.23 maf; if Mead < 1075 feet, balance contents with a min/max release of 7.0 and 9.0 maf	
3575		9.52
3525	7.48 maf 8.23 maf if Mead < 1025 f	5.93
	Balance contents with a min/max release of 7.0 and 9.5 maf	
3370		0

Lake Powell Equalization Elevation Table

In each of the following years, the Lake Powell Equalization Elevation will be as follows:

Year	Elevation (feet)
2008	3636
2009	3639
2010	3642
2011	3643
2012	3645
2013	3646
2014	3648
2015	3649
2016	3651
2017	3652
2018	3654
2019	3655
2020	3657
2021	3659
2022	3660
2023	3662
2024	3663
2025	3664

1. Equalization: In years when Lake Powell content is projected on January 1 to be at or above the elevation stated in the Lake Powell Equalization Elevation Table, an amount of water will be released from Lake Powell to Lake Mead at a rate greater than 8,230,000 acre-feet per year to the extent necessary to equalize storage in the two reservoirs, or otherwise to release 8,230,000 acre-feet from Lake Powell.

2. Upper Elevation Balancing: In years when Lake Powell content is projected on January 1 to be below the elevation stated in the Lake Powell Equalization Elevation Table and at or above 3575 ft., the Secretary shall release 8,230,000 acre-feet from Lake Powell if the projected elevation of Lake Mead is at or above 1075 ft. If the projected elevation of Lake Mead is below 1075 ft., the Secretary shall balance the contents of Lake Mead and Lake Powell, but shall release no more than 9,000,000 acre-feet and no less than 7,000,000 acre-feet from Lake Powell.

3. Mid-Elevation Releases: In years when Lake Powell content is projected on January 1 to be below 3575 ft. and at or above 3525 ft., the Secretary shall release 7,480,000 acre-feet from Lake Powell if the projected elevation of Lake Mead is at or above 1025 ft. If the projected elevation of Lake Mead is below 1025 ft., the Secretary shall release 8,230,000 acre-feet from Lake Powell.

4. Lower Elevation Balancing: In years when Lake Powell content is projected on January 1 to be below 3525 ft., the Secretary shall balance the contents of Lake Mead and Lake Powell, but shall release no more than 9,500,000 acre-feet and no less than 7,000,000 acre-feet from Lake Powell.

Coordinated Operation of Lakes Powell and Mead as described herein will be presumed to be consistent with the Section 602(a) storage requirement contained in the Colorado River Basin Project Act.

The objective of the operation of Lakes Powell and Mead as described herein is to avoid curtailment of uses in the Upper Basin, minimize shortages in the Lower Basin and not adversely affect the yield for development available in the Upper Basin.

The August 24-month study projections for the January 1 system storage and reservoir water surface elevations, for the following year, would be used to determine the applicability of the coordinated operation of Lakes Powell and Mead.

Section 3.   Determination of Lake Mead Operation during the Interim Period

A. Interim Surplus Guidelines

1. The Basin States recommend that the Secretary continue to implement the Interim Surplus Guidelines (ISG) except as modified by this proposal, including the following:

Appendix A

        a. Partial Domestic Surplus would be discontinued upon issuance of the Record Of Decision ("ROD"); and

        b. The ISG effective period would be extended through December 31, 2025.

2. During the years 2017 through 2025 the Secretary shall distribute Domestic Surplus water:

        a. For use by MWD, 250,000 acre-feet per year in addition to the amount of California's basic apportionment available to MWD.

        b. For use by SNWA, 100,000 acre-feet per year in addition to the amount of Nevada's basic apportionment available to SNWA.

        c. For use in Arizona, 100,000 acre-feet per year in addition to the amount of Arizona's basic apportionment available to Arizona contractors.

B. Flood Control Surplus

In years in which the Secretary makes space building or flood control releases pursuant to the Field Working Agreement, the Secretary shall determine a Flood Control Surplus for the remainder of that year or the subsequent year as specified in Section 7 of the ISG. In such years, releases will be made to satisfy all beneficial uses within the United States, including unlimited off-stream banking. Intentionally Created Surplus credits, as defined herein, would be reduced by the amount of any flood control release, if necessary until no credits are remaining. Under current practice, surplus declarations under the Treaty for Mexico are declared when flood control releases are made. Operation under a Flood Control Surplus does not establish any determination relating to implementation of the Treaty, including any potential changes in approach relating to surplus declarations under the Treaty. Such determinations must be addressed in a bilateral fashion with the Republic of Mexico.

C. Quantified Surplus
   (70R Strategy)

In years when the Secretary determines that water should be released for beneficial consumptive use to reduce the risk of potential reservoir spills based on the 70R Strategy, the Secretary shall determine and allocate Quantified Surplus sequentially as follows:

1. Establish the volume of the Quantified Surplus. For the purpose of determining the existence, and establishing the volume, of Quantified Surplus, the Secretary would not consider the volume of Intentionally Created Surplus credits, as defined herein.

2. Allocate and distribute the Quantified Surplus 50% to California, 46% to Arizona and 4% to Nevada, subject to 3. through 5. that follow.

3. Distribute California's share first to meet basic apportionment demands and MWD's demands. Then distribute to California Priorities 6 and 7 and other surplus contracts. Distribute Nevada's share first to meet basic apportionment demands and SNWA's demands. Distribute Arizona's share to surplus demands in Arizona including off stream banking and interstate banking demands. Arizona, California and Nevada agree that Nevada would get first priority for interstate banking in Arizona.

4. Distribute any unused share of the Quantified Surplus in accordance with Section 1, Allocation of Unused Basic Apportionment Water Under Article II (B)(6).

5. Determine whether MWD, SNWA and Arizona have received the amount of water they would have received under Section 3 D of this proposal, Domestic Surplus, if a Quantified Surplus had not been declared. If they have not, then determine and meet all demands provided for in Section 3 D, Domestic Surplus.

D. Domestic Surplus

In years when Lake Mead elevation is projected on January 1 to be above 1145 ft and below 70R Strategy elevation determination, the Secretary would determine a Domestic Surplus in accordance with Section 2(B)(2) of the ISG between the effective date of the ROD and December 31, 2016 and in accordance with Section 3(A) (2) of this proposal between January 1, 2017 and December 31, 2025.

E. Normal Conditions

In years when Lake Mead elevation is projected on January 1 to be above elevation 1075 ft. and below 1145 ft., the Secretary would determine a normal operating condition. In any year when Lake Mead elevations are in this range, the Secretary may determine that Intentionally Created Surplus ("ICS") as described in Section 4 of this proposal is available. ICS credits may then be delivered pursuant to the provisions of Section 4.

F. Shortage Conditions

Shortages would be implemented in the Lower Division States and Mexico under the following conditions and in the following manner:

1. 400,000 acre foot shortage: In years when Lake Mead content is projected on January 1 to be at or below elevation 1075 ft. and at or above 1050 ft., a quantity of 400,000 acre-feet shall not be released or delivered in the Lower Division States and Mexico.

2. 500,000 acre foot shortage: In years when Lake Mead content is projected on January 1 to be below elevation 1050 ft. and at or above 1025 ft. a quantity of 500,000 acre-feet shall not be released or delivered in the Lower Division States and Mexico.

Appendix A

3. 600,000 acre foot shortage: In years when Lake Mead content is projected on January 1 to be below 1025 ft., a quantity of 600,000 acre-feet shall not be released or delivered in the Lower Division States and Mexico.

4. The three conditions described above are illustrated in Figure 2.

Figure 2

Lake Mead Step Shortage		
Mead Elevation (ft)	Stepped Shortage	Mead Live Storage
1075 to 1050	400 kaf	9.37 to 7.47 maf
<1050 to 1025	500 kaf	7.47 to 5.80 maf
<1025 to 1000	600 kaf	5.80 to 4.33 maf
<1000	Increased reductions to be consistent with consultation(s)	<4.33 maf

5. The United States, through the appropriate mechanisms, should implement a shortage pursuant to Article 10 of the 1944 Treaty in any year in which the Secretary has declared that a shortage condition exists pursuant to Art. II(B)(3) of the Decree. The total quantity of water that will not be released or delivered to Mexico shall be based on Lower Basin water deliveries during normal water supply conditions. The proportion of the shortage that shall be borne by Mexico will be 17% (1.5 maf / 9 maf x 100% = 17%).

6. Arizona and Nevada will share shortages based on a shortage sharing agreement. In the event that no agreement has been reached, Arizona and Nevada will share shortages in accordance with the 1968 Colorado River Basin Project Act, the Decree, other existing law as applicable, and the Interstate Banking Agreement between Arizona and Nevada parties.

7. Whenever Lake Mead reaches elevation 1025 ft., the Secretary will consult with the States to determine whether Colorado River hydrologic conditions, together with the delivery of 8.4 million acre-feet of Colorado River water to Lower Basin users and Mexico, will cause the elevation of Lake Mead to fall below 1000 ft. Upon such a determination, the Secretary shall consult with the states to discuss further measures that may be undertaken to avoid or reduce further increases in shortage determinations. If increased reductions are required, the Secretary shall implement the reductions consistent with the law of the river.

8. The States will evaluate factors at critical elevations that may avoid shortage determinations as reservoir elevations approach critical thresholds. The States may provide operational recommendations surrounding the critical elevations at some later date.

Section 4. System Efficiency, Extraordinary Conservation and Augmentation Projects

The States propose that the Secretary develop a policy and accounting procedure concerning augmentation, extraordinary conservation, and system efficiency projects, including specific extraordinary conservation projects, tributary conservation projects, introduction of non-Colorado River System water, system efficiency improvements and exchange of non-Colorado River System water. The accounting and recovery process would be referred to as "Intentionally Created Surplus" consistent with the concept that the States will take actions to augment storage of water in the Lower Colorado River Basin. The water would be distributed pursuant to Section II(B)(2) of the Decree and forbearance agreements between the States. The ICS credits may not be created or released without such forbearance agreements.

   A. The purposes of the Lake Mead Intentionally Created Surplus ("ICS") program are to:

      1. Help avoid shortages to the Lower Basin. For the purposes of determining calendar year declarations of Domestic Surplus, Normal and Shortage conditions, any ICS credits would be considered system water;

      2. Benefit both Lake Mead and Lake Powell; and

      3. Increase the surface elevations of both Lakes Powell and Mead to higher levels than would have otherwise occurred.

   B. Extraordinary Conservation Storage Credits

      1. Users of Colorado River water may create ICS credits through extraordinary conservation under the following conditions:

         a. A Boulder Canyon Project Act Section 5 Contractor ("Contractor") shall repay all outstanding system payback obligations before it can create ICS credits.

         b. ICS credits can only be created if such water could have otherwise been beneficially used.

         c. A Contractor notifies Reclamation by September 15 of the amount of ICS credits it wishes to create for the subsequent year.

      2. ICS credits may be created only through extraordinary conservation activities. These activities include:

         a. Fallowing of land that currently is, historically was, and otherwise would have been in the next year, irrigated.

         b. Canal lining programs

         c. Desalination programs

Appendix A

      d.   Extraordinary conservation programs existing as of January 1, 2006

      e.   Other extraordinary conservation measures as agreed upon by the States

3. If conditions during the year change due to weather or other unforeseen circumstances, a Contractor may request a mid-year modification of its water order to reduce the amount of ICS credits created during that year. A Contractor cannot increase the amount of ICS credits it had previously scheduled to create during the year.

4. Any ICS credits would be used first to offset any overrun for that year or future year(s).

5. The maximum amount of ICS credits that can be created during any year through extraordinary conservation is limited to each state as listed below.

      a.   California: 400,000 acre-feet per year

      b.   Nevada: 125,000 acre-feet per year

      c.   Arizona: 100,000 acre-feet per year

6. The maximum cumulative amount of ICS credits created through extraordinary conservation that would be available at any one time is:

      a.   1,500,000 acre-feet for California;

      b.   300,000 acre-feet for Nevada; and

      c.   300,000 acre-feet for Arizona.

7. No category of surplus water can be used to create ICS credits.

8. At the time the ICS credits are created by extraordinary conservation, the Contractor will dedicate 5% of the ICS credits to the system on a one-time basis to provide a water supply benefit to the system. Additionally, ICS credits will be subject to annual evaporation loss (estimated to be no more than 3% annually) during each year in which no shortage has been declared. The Secretary will not assess any other charge for creating ICS credits.

9. Contractors that have created ICS credits may recover them under the following conditions:

      a.   A Contractor may request delivery of ICS credits it has created at the time it submits its annual water order for the following year. The ICS credits would be added to the Contractor's approved water order for that year upon approval by Reclamation.

b. The amount of ICS credits that may be recovered by California in any one year is limited to 400,000 acre-feet, by Nevada 300,000 acre-feet and Arizona 300,000 acre-feet; provided that the May 1, 24-month study for that year does not indicate that a shortage condition would be declared in the current or succeeding year.

c. If extraordinary weather conditions or water emergencies occur, a Contractor may request that Reclamation increase its use of ICS credits for that year.

d. A Contractor may request to reduce its use of ICS credits during the year for any reason, including reduction in water demands.

e. If Reclamation releases water for flood control purposes, ICS credits shall be reduced on a pro-rata basis among all holders of ICS credits-- if necessary until no credits remain. In determining the amount of Quantified Surplus, Reclamation shall not consider the volume of ICS credits that will be available.

10. Contractors may begin to create ICS through extraordinary conservation 1) beginning in 2006 as a pilot program (which may be lost if the Secretary does not adopt an extraordinary conservation program as part of the Coordinated Operation of Lakes Powell and Mead) or 2) after adoption of the Coordinated Operation for Lakes Powell and Mead until 2025. Any ICS credits under this program remaining at the end of the program would remain available for recovery for up to 10 years following termination of the Program.

C. Tributary Conservation

The Secretary should develop procedures in consultation with the States that would permit Contractors to purchase and fallow annual or permanent water rights on tributaries within the Lower Division States that have been used for a significant period of years and were created prior to Congress' adoption of the Boulder Canyon Project Act that, when retired, and verified by the Secretary, contribute water to the Colorado River mainstream for diversion by the Contractor. The water recovered by the Contractor may be used for municipal and industrial purposes only. This water would be in addition to the State's basic apportionment and would be available during declared shortages.

It is intended that the water would be taken on a real-time basis and that not more than 95% of such water will be recovered; however, if storage were required, such stored water would be subject to all provisions applicable to ICS credits created through extraordinary conservation.

Appendix A

D. System Efficiency Projects

A Contractor may make contributions of capital to the Secretary for use in Secretarial projects designed to realize efficiencies that save water that would otherwise be lost from the Colorado River System in the United States. The Secretary in consultation with the States will identify system efficiency projects, terms for capital participation in such projects, and types and amounts of benefits the Secretary would provide in consideration of non-federal capital contributions to system efficiency projects, including a portion of the water saved by the project. Water made available to Contractors by the Secretary would be considered Intentionally Created Surplus. System efficiency projects are only intended to provide temporary water supplies and would not be available for permanent use.

Benefits to the total water available within the Colorado River System in the United States should be substantial, taking into account any benefit provided to any non-federal capital contributor. In those cases in which benefits are provided to a non-federal capital contributor in the form of a portion of the water saved by the system efficiency project, the water provided to the capital contributor should be characterized as Colorado River surplus water intentionally created by the system efficiency project. The ICS credits should be provided to the capital contributor pursuant to its BCPA § 5 surplus contract. The Secretary should first obtain the waiver or forbearance of any other BCPA § 5 surplus contractor(s) that may possess any right to the delivery of the same water, so that the Secretary may deliver it to the capital contributor pursuant to Article II (B)(6) of the Decree. The ICS credits should be provided to the capital contributor on a predetermined schedule of annual deliveries for a period of years as agreed by the Secretary and Contractor. The ICS credits would not be stored, and therefore would not spill from system reservoirs. Delivery of ICS credits during shortage conditions will be determined on a project-by-project basis.

E. Introduction and Recovery of Non-Colorado River System Water

The Secretary should develop procedures, in consultation with the States, that would prospectively allow non-Colorado River System water in a Lower Division State to be introduced into, conveyed through, and diverted from system reservoirs, or otherwise through the Colorado River System. The non-Colorado River System water may be introduced either (1) directly from the non-Colorado River System source, or (2) as effluent resulting from use of the non-Colorado River System water in the introducing entity's service area, assuming water quality concerns are adequately addressed by the Contractor introducing the water. This water is in addition to a state's basic apportionment and may be used during declared shortages.

Contractors proposing to introduce, convey and recover such non-Colorado River System water should make sufficient arrangements, contractual or otherwise, with the Secretary so as to guarantee that any such action causes no harm to the Secretary's management of the Colorado River System. Such arrangements would provide that the introduction, conveyance and recovery of such water be done pursuant to appropriate permits or other authorizations as required by state law, that the actual amount of water introduced, conveyed and recovered would be reported to the Secretary on an annual basis, and that no more than 95% of such water introduced will be recovered. The non-Colorado River System water would be intended to be taken on a real-time basis, and hence would not spill from system reservoirs. However, if storage were required such stored water would be subject to all provisions applicable to ICS created through extraordinary conservation. Any agreements made with the Secretary to introduce and recover this water will survive the termination of the Coordinated Operations of Lakes Powell and Mead.

Weather modification projects should be pursued as a means of augmenting Colorado River System water supplies. However, increases in water supply that result from weather modification projects are not included within the projects defined in this Section and would not create any additional supply for a Contractor or State that engages in a weather modification project.

Section 5.   Non-Colorado River System Water Exchanges

Contractors in Arizona, California, or Nevada may secure an additional water supply by funding the development of a non-Colorado River System water supply in one Lower Division State for use in another State by exchange. The new water supply developed would be consumptively used in the State in which it was developed by a Contractor and that Contractor would intentionally reduce its consumptive use of Colorado River water. This would allow the Contractor(s) in the other Lower Division State(s) that provided the funding to consumptively use the Colorado River water that was intentionally unused through an agreement with the Secretary of the Interior. Through the cooperation of the International Boundary and Water Commission, United States and Mexico, similar agreements could be established by which non-Colorado River System water supplies in Mexico could be developed for use in the United States by exchange.

It could be necessary for a State or other lower priority Contractors in the State in which consumptive use was intentionally reduced to agree to forebear their use of such water depending on the then-existing priority system to use of Colorado River water, to avoid a claim against the water being delivered to the Contractor that funded the new water supply. As an alternative to forbearance, an offer by the Contractor developing the non-Colorado River System water to allow the lower priority Contractor to pay the cost of developing a portion or all of the non-Colorado River System water supplies to be developed, would be utilized to protect such a lower priority Contractor's position in the then-existing priority system. A refusal of an offer to pay the cost of developing a portion or all of the non-Colorado River System water supplies to be developed would constitute the lower-priority Contractor's waiver of a right to challenge the exchange.

Section 6. Accounting Mechanisms

The operating alternatives discussed in Sections 4 and 5 will require new or modified Colorado River accounting mechanisms. No specific accounting mechanism to allow these types of operations is proposed for evaluation in Reclamation's current NEPA process. However, the description and evaluation of such accounting mechanisms would provide Contractors with the assurance that if such accounting mechanism were adopted in the Record of Decision, funds spent to propose such an arrangement in the future would not be spent in vain.

Appendix A

Section 7.   Effective Period

The proposed interim operations will be in effect 30 days from the publication of the Secretary's Record of Decision in the Federal Register. The proposed interim operations will, unless subsequently modified, remain in effect through December 31, 2025 (through preparation of the 2026 AOP), subject to a formal review of their effectiveness beginning no later than 2020.

Attachment B

DRAFT

AGREEMENT

The [name parties] hereby enter into this Agreement effective as of _____.

RECITALS

A. Parties.

   1. Arizona

      a. The Arizona Department of Water Resources, through its Director, is the successor to the signatory agency of the State for the 1922 Colorado River Compact, and the 1944 Contract for Delivery of Water with the United States, both authorized and ratified by the Arizona Legislature, A.R.S. §§ 45-1301 and 1311. Pursuant to A.R.S. §§ 45-107, the Director is authorized and directed, subject to the limitations in A.R.S. §§ 45-106, for and on behalf of the State of Arizona, to consult, advise and cooperate with the Secretary of the Interior of the United States with respect to the exercise by the Secretary of Congressionally authorized authority relative to the waters of the Colorado River (including but not limited to the Boulder Canyon Project Act, 43 U.S.C. § 617, and the 1968 Colorado River Basin Project Act, 43 U.S.C. § 1501) and with respect to the development, negotiation and execution of interstate agreements. Additionally, under A.R.S. § 45-105(A)(9), the Director is authorized to "prosecute and defend all rights, claims and privileges of this state respecting interstate streams."

      b. Under A.R.S. § 11-951 *et. seq.*, the Director is authorized to enter into Intergovernmental Agreements with other public agencies, which includes another state; departments, agencies, boards and commissions of another state; and political subdivisions of another state.

   2. California. The chairman of the Colorado River Board of California, acting as the Colorado River Commissioner pursuant to California Water Code section 12525, has the authority to exercise on behalf of California every right and power granted to California by the Boulder Canyon Project Act, and to do and perform all other things necessary or expedient to carry out the purposes of the Colorado River Board.

   3. Colorado

      a. Section 24-1-109, Colorado Revised Statutes (2005) provides that "Interstate compacts authorized by law shall be administered under the direction of the office of the governor." This includes the Colorado River Compact and the Upper Colorado River Basin Compact. Section 37-60-109 provides that "the governor from time to time, with approval of the

Appendix A

DRAFT

board, shall appoint a commissioner, who shall represent the state of Colorado upon joint commissions to be composed of commissioners representing the state of Colorado and another state or other states for the purpose of negotiating and entering into compacts or agreements between said states…" By Executive Order _____, issued _____, 2006, attached hereto as Exhibit _____ and incorporated herein by reference, the Governor appointed Upper Colorado River Commissioner Scott Balcomb to represent the State of Colorado.

b. Section 37-60-106, subsections (e) and (i), C.R.S. (2005), authorize the Colorado Water Conservation Board to "cooperate with the United States and the agencies thereof, and with other states for the purpose of bringing about the greater utilization of the water of the state of Colorado and the prevention of flood damages," and "to confer with and appear before the officers, representatives, boards, bureaus, committees, commissions, or other agencies of other states, or of the federal government, for the purpose of protecting and asserting the authority, interests, and rights of the state of Colorado and its citizens with respect to the waters of the interstate streams in this state." By resolution dated _____, attached hereto as Exhibit __, and incorporated herein by reference, the Colorado Water Conservation Board authorized and directed its Director to negotiate with and enter into agreements with other state entities within the Colorado River Basin.

4. Nevada

a. The Colorado River Commission of the State of Nevada (CRCN) is an agency of the State of Nevada, authorized generally by N.R.S. §§ 538.041 and 538.251. CRCN is authorized by N.R.S. § 538.161 (6), (7) to enter into this Agreement. The CRCN, in furtherance of the State of Nevada's responsibility to promote the health and welfare of its people in Colorado River matters, makes this Agreement to supplement the supply of water in the Colorado River which is available for use in Nevada, augment the waters of the Colorado River, and facilitate the more flexible operation of dams and facilities by the Secretary of the Interior of the United States. The Chairman of the Commission, signatory hereto, serves as one of the Governor's representatives as contemplated by Section 602(b) of the 1968 Colorado River Basin Project Act, 43 U.S.C. § 1552(b) and the Criteria for Coordinated Long-Range Operation of Colorado River Reservoirs Pursuant to the Colorado River Basin Project Act.

b. The Southern Nevada Water Authority (SNWA) is a Nevada joint powers agency and political subdivision of the State of Nevada, created by agreement dated July 25, 1991, as amended November 17,1994 and January 1,1996, pursuant to N.R.S. §§ 277.074 and 277.120. SNWA is authorized by N.R.S. § 538.186 to enter into this Agreement and, pursuant

DRAFT

to its contract issued under section 5 of the Boulder Canyon Project Act of 1928, SNWA has the right to divert "supplemental water" as defined by NRS § 538.041 (6). The General Manager of the SNWA, signatory hereto, serves as one of the Governor's Representatives as contemplated by Section 602(b) of the 1968 Colorado River Basin Project Act, 43 U.S.C. § 1552(b) and the Criteria for Coordinated Long-Range Operation of Colorado River Reservoirs Pursuant to the Colorado River Basin Project Act.

5. New Mexico. Pursuant to NMSA 1978, 72-14-3, the New Mexico Interstate Stream Commission is authorized to investigate water supply, to develop, to conserve, to protect and to do any and all other things necessary to protect, conserve and develop the waters and stream systems of the State of New Mexico, interstate or otherwise. The Interstate Stream Commission also is authorized to institute or cause to be instituted in the name of the state of New Mexico any and all negotiations and/or legal proceedings as in its judgment are necessary. By Resolution dated _____, the Interstate Stream Commission authorizes the execution of this Agreement.

6. Utah. The Division of Water Resources (DWR) is the water resource authority for the State of Utah. Utah Code Ann. § 73-10-18. The Utah Department of Natural Resources Executive Director (Department), with the concurrence of the Utah Board of Water Resources (Board), appoints the DWR Director (Director). § 63-34-6(1). The Board makes DWR policy. § 73-10-1.5. The Board develops, conserves, protects, and controls Utah waters, § 73-10-4(4),(5), and, in cooperation with the Department and Governor, supervises administration of interstate compacts, § 73-10-4, such as the Colorado River Compact, §§ 73-12a-1 through 3, and the Upper Colorado River Basin Compact, § 73-13-10. The Board, with Department and Gubernatorial approval, appoints a Utah Interstate Stream Commissioner, § 73-10-3, currently the DWR Director, to represent Utah in interstate conferences to administer interstate compacts. §§ 73-10-3 and 73-10-4. These delegations of authority authorize the Utah Interstate Stream Commissioner/DWR Director to sign this document. He acts pursuant to a Board resolution, acknowledged by the Department, dated _____, attached hereto as Exhibit __, and incorporated herein by reference.

7. Wyoming. Water in Wyoming belongs to the state. WYO. CONST. Art. 8 ' 1. The Wyoming State Engineer is a constitutionally created office and is Wyoming's chief water official with general supervisory authority over the waters of the state. WYO. CONST. Art. 8 ' 5. The Wyoming legislature conferred upon Wyoming officers the authority to cooperate with and assist like authorities and entities of other states in the performance of any lawful power, duty, or authority. WYO. STAT. ANN. ' 16-1-101 (LEXISNEXIS 2005). Wyoming and its State Engineer represent the rights and interests of all Wyoming appropriators with respect to other states. *Wyoming v. Colorado*,

3

Appendix A

DRAFT

286 U.S. 494 (1922). *See Hinderlider v. La Plata River & Cherry Creek Ditch Co.*, 304 U.S. 92 (1938). In signing this Agreement, the State Engineer intends that this Agreement be mutually and equally binding between the Parties.

B. Background

1. Federal law and practice (including Section 602(b) of the 1968 Colorado River Basin Project Act, 43 U.S.C. § 1552(b), and the Criteria for Coordinated Long-Range Operation of Colorado River Reservoirs Pursuant to the Colorado River Basin Project Act), contemplate that in the operation of Lakes Powell and Mead, the Secretary of the Interior consults with the States through Governors' Representatives, who represent the Governors and their respective States. Through this law and practice, the Governors' Representatives have in the past reached agreements among themselves and with the Secretary on various aspects of Colorado River reservoir operation. This Agreement is entered into in furtherance of this law and practice.

2. On January 16, 2001, the Secretary adopted Colorado River Interim Surplus Guidelines (ISG) based on an alternative prepared by the Colorado River Basin States, for the purposes of determining annually the conditions under which the Secretary would declare the availability of surplus water for use within the states of Arizona, California and Nevada in accordance with and under the authority of the Boulder Canyon Project Act of 1928 (45 Stat. 1057) and the Decree of the United States Supreme Court in *Arizona v. California*, 376 U.S. 340 (1964). The ISG are effective through calendar year 2015 (through preparation of the 2016 Annual Operating Plan).

3. In the years following the adoption of the ISG, drought conditions in the Colorado River Basin caused a significant reduction in storage levels in Lakes Powell and Mead, and precipitated discussions by and among the Parties, and between the Parties and the United States through the Department of the Interior and the Bureau of Reclamation. The Parties recognize that the Upper Division States have not yet developed their full apportionment under the Colorado River Compact. Although the Secretary has not imposed any shortage in the Lower Basin, the Parties also recognize that with additional Upper Basin development and in drought conditions, the Lower Division States may be required to suffer shortages in deliveries of water from Lake Mead. Therefore, these discussions focused on ways to improve the management of water in Lakes Powell and Mead so as to enhance the protection afforded to the Upper Basin by Lake Powell, and to delay the onset and minimize the extent and duration of shortages in the Lower Basin.

4. Shortages in the Lower Basin will also trigger shortages in the delivery of water to Mexico pursuant to the Mexican Water Treaty of 1944, February 3, 1944, U.S.-Mex., 59 Stat. 1219, T.S. 994, 3 U.N.T.S. 313.

DRAFT

5. On May 2, 2005, the Secretary announced her intent to undertake a process to develop Lower Basin shortage guidelines and explore management options for the coordinated operation of Lakes Powell and Mead. On June 15, 2005, the Bureau of Reclamation published a notice in the *Federal Register*, announcing its intent to implement the Secretary's direction. The Bureau of Reclamation has proceeded to undertake scoping and develop alternatives pursuant to the National Environmental Policy Act (the NEPA Process), which the Parties anticipate will form the basis for a ROD to be issued by the Secretary by December 2007.

6. On August 25, 2005, the Governors' Representatives for the seven Colorado River Basin States wrote a letter to the Secretary expressing conceptual agreement in the development and implementation of three broad strategies for improved management and operation of the Colorado River: Coordinated Reservoir Management and Lower Basin Shortage Guidelines; System Efficiency and Management; and Augmentation of Supply.

7. On February 3, 2006, the Governors' Representatives transmitted to the Secretary their recommendation for the scope of the NEPA Process, which refined many of the elements outlined in the August 25, 2005 letter.

8. At the request of the Secretary, the Parties have continued their discussions relative to the areas of agreement outlined in the letters of August 25, 2005 and February 3, 2006.

9. In furtherance of the letters of August 25, 2005 and February 3, 2006, the Parties have reached agreement to take additional actions for their mutual benefit, which are designed to augment the supply of water available for use in the Colorado River System and improve the management of water in the Colorado River.

C. Purpose. The Parties intend that the actions by them contemplated in this Agreement will: improve cooperation and communication among them; provide additional security and certainty in the water supply of the Colorado River System for the benefit of the people served by water from the Colorado River System; and avoid circumstances which could otherwise form the basis for claims or controversies over interpretation or implementation of the Colorado River Compact and other applicable provisions of the law of the river.

AGREEMENT

In consideration of the above recitals and the mutual covenants contained herein, and other good and valuable consideration, the receipt and sufficiency of which is hereby acknowledged, the Parties agree as follows:

1. Recitals. The Recitals set forth above are material facts that are relevant to and form the basis for the agreements set forth herein.

Appendix A

DRAFT

2. <u>Definitions</u>. As used in this Agreement, the following terms have the following meanings:

   A. <u>Colorado River System</u>. This term shall have the meaning as defined in the Colorado River Compact.

   B. <u>ISG</u>. The Colorado River Interim Surplus Guidelines adopted by the Secretary on January 16, 2001.

   C. <u>NEPA Process</u>. The decision-making process pursuant to the National Environmental Policy Act, 42 U.S.C. §§ 4321 through 47, beginning with the Bureau of Reclamation's Notice to SolicitComments and Hold Public Meetings, 70 Fed. Reg. 34794 (June 15, 2005) and culminating in a Record of Decision.

   D. <u>Party or Parties</u>. Any party or parties to this Agreement.

   E. <u>Parties' Recommendation</u>. The Seven Basin States' Preliminary Proposal Regarding Colorado River Interim Operations, a copy of which is attached hereto and incorporated herein by this reference, presented by the Parties to the Secretary in furtherance of the States' letters of August 25, 2005 and February 3, 2006, and any modification of the Parties' Recommendation adopted by the Parties pursuant to this Agreement.

   F. <u>ROD</u>. The Record of Decision anticipated to be issued by the Secretary after completion of NEPA Process, pursuant to her letter of May 2, 2005, and the Notice published in the Federal Register on September 30, 2005, 70 Fed. Reg. 57322.

   G. <u>Secretary</u>. The Secretary of the Interior or the Bureau of Reclamation, as applicable.

   H. <u>State or States</u>. Any of the states of Arizona, California, Colorado, Nevada, New Mexico, Utah or Wyoming, as context requires.

3. <u>Support for Parties' Recommendation</u>. After considering a number of alternatives, each Party has determined that the Parties' Recommendation is in the best interests of that Party, and promotes the health and welfare of that Party and of the Colorado River Basin States. In the NEPA Process, the Parties shall support the Secretary's adoption of the Parties' Recommendation in a ROD. If during the course of the NEPA Process any new information becomes available which causes any Party, in its sole and absolute discretion, to reassess any provision of the Parties' Recommendation, that Party shall immediately notify all other Parties in writing. The Parties shall jointly confer and, if they agree to any modification of the Parties' Recommendation, shall consult with the Secretary to advise her of such modification and request the adoption thereof in the ROD. If after such conference and consultation it is apparent there is an

DRAFT

irreconcilable conflict between the Parties as to such modification, then any Party may upon written notice to the other Parties withdraw from this Agreement, and in such event this Agreement shall no longer be effective or binding upon such withdrawing Party. All withdrawing Parties hereby reserve all rights upon withdrawal from this Agreement to take such actions, including support of or challenges to the ROD, as they in their sole and absolute discretion deem necessary or appropriate. In the event of the withdrawal of any one or more Parties from this Agreement, this Agreement shall continue in full force and effect as to the remaining Parties. The remaining Parties may confer to determine whether to continue this Agreement in effect, to amend this Agreement, or to terminate this Agreement. In the event of termination, all Parties shall be relieved from the terms hereof, and this Agreement shall be of no further force or effect.

4. <u>ROD Consistent with the Parties' Recommendation</u>. In the event the Secretary adopts a ROD in substantial conformance with the Parties' Recommendation, the Parties shall take all necessary actions to implement the terms of the ROD, including the approval and execution of agreements necessary for such implementation.

5. <u>ROD Inconsistent with the Parties' Recommendation</u>. In the event the Secretary adopts a ROD that any Party, in its sole and absolute discretion, determines is not in substantial conformance with the Parties' Recommendation, such Party shall immediately notify all other Parties of such determination in writing. The Parties shall jointly confer, and consult with the Secretary as necessary, in order to determine whether the ROD is in substantial conformance with this Agreement, or whether any action, including the amendment of this Agreement, may resolve such concern. If after such conference and consultation it is apparent there is an irreconcilable conflict between the ROD and the concerns of such Party, then such Party may upon written notice to the other Parties withdraw from this Agreement, and in such event this Agreement shall no longer be effective or binding upon such withdrawing Party. All withdrawing Parties hereby reserve all rights upon withdrawal from this Agreement to take such actions, including support of or challenges to the ROD, as they in their sole and absolute discretion deem necessary or appropriate. In the event of the withdrawal of any one or more Parties from this Agreement, this Agreement shall continue in full force and effect as to the remaining Parties. The remaining Parties may confer to determine whether to continue this Agreement in effect, to amend this Agreement, or to terminate this Agreement. In the event of termination, all Parties shall be relieved from the terms hereof, and this Agreement shall be of no further force or effect.

6. <u>Additions to the ROD</u>. The Parties hereby request that the Secretary recognize the specific provisions of this Agreement as part of the NEPA Process and, if appropriate, include in the ROD specific provisions that reference this Agreement as a basis for the ROD. The Parties also hereby request that the Secretary include in the ROD specific provision that the Secretary will first consult with all the States, through their designated Governor's Representatives, before making any substantive modification to the ROD. Finally, the Parties hereby request that the Secretary include in the ROD specific provision that upon a request by any State for modification of the ROD, or upon any request by any State to resolve any claim or controversy arising under this Agreement or

7

Appendix A

DRAFT

under the operations of Lakes Powell and Mead pursuant to the ROD, the ISG, or any other applicable provision of federal law, regulation, criteria, policy, rule or guideline, the Secretary shall invite all of the Governors, or their designated representatives, to consult with the Secretary in an attempt to resolve such claim or controversy by mutual agreement.

7. <u>Consultation on Operations</u>. After the Secretary commences operating Lakes Powell and Mead pursuant to the ROD, the Parties shall confer among themselves as necessary, but at least annually, to assess such operations. Any Party may request consultation with the other Parties on a proposed adjustment or modification of such operations, based on changed circumstances, unanticipated conditions, or other factors. Upon such request, the Parties shall in good faith confer to resolve any such issues, and based thereon may request consultation by the States with the Secretary on adjustments to or modifications of operations under the ROD. In any event, the Parties shall confer before December 31, 2020, to determine whether to extend this Agreement and recommend that the Secretary continue operations under the ROD for an additional period, or modify this Agreement and recommend that the Secretary modify operations under the ROD, or terminate this Agreement and recommend that the Secretary not continue operations under the ROD after the expiration thereof.

8. <u>Development of System Augmentation</u>. The Parties agree to diligently pursue system augmentation within the Colorado River System including but not limited to the determination of the feasibility of projects to increase precipitation in the basin or to augment available supplies through desalination. Additionally, the Parties agree to cooperatively pursue an interim water supply of at least a cumulative amount of 280,000 acre-feet for use in Nevada while long-term augmentation projects are being pursued. It is anticipated that this interim water supply will be made available in return for Nevada's funding of the Drop 2 Reservoir currently proposed for construction by the Bureau of Reclamation. Annual recovery of this interim water supply by Nevada will not exceed 40,000 acre-feet. All water available to Nevada in consideration for funding the Drop 2 Reservoir would remain available during all shortage conditions declared by the Secretary.

In consideration of the Parties' diligent pursuit of long-term augmentation and the availability of the interim water supply, the Southern Nevada Water Authority (SNWA) agrees that it will withdraw right-of-way Application No. N-79203 filed with the Bureau of Land Management on October 1, 2004 for the purpose of developing Permit No. 58591 issued by the Nevada State Engineer in Ruling No. 4151.

The SNWA will not re-file such right-of-way application or otherwise seek to divert the water rights available under Permit No. 58591 from the Virgin River prior to 2014 so long as Nevada is allowed to utilize its pre-Boulder Canyon Project Act Virgin and Muddy River rights in accordance with section 4(C) of the Parties' Recommendation in the form forwarded to the Secretary on February 3, 2006, and the interim water supply made available to Nevada is reasonably certain to remain available. The SNWA will not re-file such right-of-way application or otherwise seek to divert the water rights available

DRAFT

under Permit No. 58591 from the Virgin River after 2014 so long as diligent pursuit of system augmentation is proceeding to provide Nevada an annual supply of 75,000 acre-feet by the year 2020. Prior to re-filing any applications with the Bureau of Land Management, SNWA and Nevada will consult with the other Basin States.

This agreement is without prejudice to any Party's claims, rights or interests in the Virgin or Muddy River systems.

9. <u>Consistency with Existing Law</u>. The Parties' Recommendation is consistent with existing law. The Parties expressly agree that the storage of water in and release of water from Lakes Powell and Mead pursuant to a ROD issued by the Secretary in substantial conformance with the Parties' Recommendation and this Agreement, and any agreements, rules and regulations adopted by the Secretary or the parties to implement such ROD, shall not constitute a violation of Article III(a)-(e) inclusive of the Colorado River Compact, or Sections 601 and 602(a) of the Colorado River Basin Project Act of 1968 (43 U.S.C. §§ 1551 and 1552(a)), and all applicable rules and regulations promulgated thereunder.

10. <u>Resolution of Claims or Controversies</u>. The Parties recognize that litigation is not the preferred alternative to the resolution of claims or controversies concerning the law of the river. In furtherance of this Agreement, the Parties desire to avoid litigation, and agree to pursue a consultative approach to the resolution of any claim or controversy. In the event that any Party becomes concerned that there may be a claim or controversy under this Agreement, the ROD, Article III(a)-(e) inclusive of the Colorado River Compact, or Sections 601 and 602(a) of the Colorado River Basin Project Act of 1968 (43 U.S.C. §§ 1551 and 1552(a)), and all applicable rules and regulations promulgated thereunder, such Party shall notify all other Parties in writing, and the Parties shall in good faith meet in order to resolve such claim or controversy by mutual agreement prior to any litigation. No Party shall initiate any judicial or administrative proceeding against any other Party or against the Secretary under Article III(a)-(e) inclusive of the Colorado River Compact, or Sections 601 and 602(a) of the Colorado River Basin Project Act of 1968 (43 U.S.C. §§ 1551 and 1552(a)), or any other applicable provision of federal law, regulation, criteria, policy, rule or guideline, and no claim thereunder shall be ripe, until such conference has been completed. In addition, all States shall comply with any request by the Secretary for consultation in order to resolve any claim or controversy. In addition, any State may invoke the provisions of Article VI of the Colorado River Compact. Notwithstanding anything in this Agreement to the contrary, the terms of this Paragraph 10 shall survive for a period of five years following the termination or expiration of this Agreement, and shall apply to any withdrawing Party after withdrawal for such period.

11. <u>Reservation of Rights</u>. Notwithstanding the terms of this Agreement and the Parties' Recommendation, in the event that for any reason this Agreement is terminated, or that the term of this Agreement is not extended, or upon the withdrawal of any Party from this Agreement, the Parties reserve, and shall not be deemed to have waived, any and all rights, including any claims or defenses, they may have as of the date hereof or as

Appendix A

DRAFT

may accrue during the term hereof, under any existing federal or state law or administrative rule, regulation or guideline, including without limitation the Colorado River Compact, the Upper Colorado River Basin Compact, the Decree in *Arizona v. California*, the Colorado River Basin Project Act of 1968, and any other applicable provision of federal law, rule, regulation, or guideline.

12. No Third-Party Beneficiaries. This Agreement is made for the benefit of the Parties. No Party to this Agreement intends for this Agreement to confer any benefit upon any person or entity not a signatory upon a theory of third-party beneficiary or otherwise.

13. Joint Defense Against Third Party Claims. In the event the Secretary adopts a ROD in substantial conformance with the Parties' Recommendation as set forth herein, they will have certain common, closely parallel, or identical interests in supporting, preserving and defending the ROD and this Agreement. The nature of this interest and the relationship among the Parties present common legal and factual issues and a mutuality of interests. Because of these common interests, the Parties will mutually benefit from an exchange of information relating to the support, preservation and defense of the ROD and this Agreement, as well as from a coordinated investigation and preparation for discussion of such interests. In furtherance thereof, in the event of any challenge by a third party as to the ROD or this Agreement (including claims by any withdrawing Party), the Parties will cooperate to proceed with reasonable diligence and to use reasonable best efforts in the support, preservation and defense thereof, including any lawsuit or administrative proceeding challenging the legality, validity or enforceability of any term of the ROD or this Agreement, and will to the extent appropriate enter into such agreements, including joint defense or common interest agreements, as are necessary therefor. Each Party shall bear its own costs of participation and representation in any such defense.

14. Reaffirmation of Existing Law. Nothing in this Agreement or the Parties' Recommendation is intended to, nor shall this Agreement be construed so as to, diminish or modify the right of any Party under existing law, including without limitation the Colorado River Compact, the Upper Colorado River Basin Compact, or the Decree in *Arizona v. California*. The Parties hereby affirm the entitlement and right of each State under such existing law to use and develop the water of the Colorado River System.

15. Term. This Agreement shall be effective as of the date of the first two signatories hereto, and shall be effective as to any additional Party as of the date of execution by such Party. Unless earlier terminated, this Agreement shall be effective for so long as the ROD and the ISG are in effect, and shall terminate upon the termination of the ROD and the ISG.

16. Authority. The persons and entities executing this Agreement on behalf of the Parties are recognized by the Parties as representing the respective States in matters concerning the operation of Lakes Powell and Mead, and as those persons and entities authorized to bind the respective Parties to the terms hereof. Each person executing this

DRAFT

Agreement has the full power and authority to bind the respective Party to the terms of this Agreement. No Party shall challenge the authority of any person or Party to execute this Agreement and bind such Party to the terms hereof, and the Parties waive the right to challenge such authority.

Appendix B

Guest Speakers at Committee Meetings

Federal

Dave Brandon, National Weather Service, Salt Lake City, Utah
Denny Fenn, Grand Canyon Monitoring and Research Center, Flagstaff, Arizona
Terry Fulp, U.S. Bureau of Reclamation, Boulder City, Nevada
Rick Gold, U.S. Bureau of Reclamation, Salt Lake City, Utah
Bob Johnson, U.S. Bureau of Reclamation, Boulder City, Nevada
Don Ostler, Upper Colorado River Commission, Salt Lake City, Utah
Kenneth Rakestraw, International Boundary and Water Commission, United States Section, El Paso, Texas
Robert H. Webb, U.S. Geological Survey, Tucson, Arizona
Robert S. Webb, National Oceanic and Atmospheric Administration, Boulder, Colorado

State[*]

Larry Anderson, Utah Division of Water Resources, Salt Lake City
Tom Carr, Arizona Department of Water Resources, Phoenix
Jeanine Jones, California Department of Water Resources, Sacramento
Rod Kuharich, Colorado Water Conservation Board, Denver
Pat Tyrrell, State of Wyoming, Laramie
John Whipple, State of New Mexico, Santa Fe

---

[*] Representatives from the State of Nevada were invited to speak with the committee but were unable to attend meetings because of scheduling conflicts.

Other

Craig Bell, Western States Water Council, Midvale, Utah
Ben Harding, Hydrosphere, Boulder, Colorado
Kathy Jacobs, University of Arizona, Tucson
Jan Matusak, Metropolitan Water District of Southern California, Los Angeles
Dave Meko, University of Arizona, Tucson
Antonio Rascón, International Boundary and Water Commission, Mexico Section, Ciudad Juárez, Chihuahua

Appendix C

Water Science and Technology Board

R. RHODES TRUSSELL, *Chair,* Trussell Technologies, Inc., Pasadena, California
MARY JO BAEDECKER, U.S. Geological Survey, Emeritus, Reston, Virginia
JOAN G. EHRENFELD, Rutgers University, New Brunswick, New Jersey
DARA ENTEKHABI, Massachusetts Institute of Technology, Cambridge, Massachusetts
GERALD E. GALLOWAY, University of Maryland, College Park
PETER GLEICK[*], Pacific Institute for Studies in Development, Environment, and Security, Oakland, California
SIMON GONZALEZ, National Autonomous University of Mexico, Mexico D.F.
CHARLES N. HAAS, Drexel University, Philadelphia, Pennsylvania
THEODORE L. HULLAR, Cornell University, Ithaca, New York
KIMBERLY L. JONES, Howard University, Washington, D.C.
KAI N. LEE, Williams College, Williamstown, Massachusetts
JAMES K. MITCHELL, Virginia Polytechnic Institute and State University, Blacksburg
CHRISTINE L. MOE[*], Emory University, Atlanta, Georgia
ROBERT PERCIASEPE, National Audubon Society, New York
LEONARD SHABMAN, Resources for the Future, Washington, D.C.
KARL K. TUREKIAN[*], Yale University, New Haven, Connecticut
HAME M. WATT, Independent Consultant, Washington, D.C.
CLAIRE WELTY, University of Maryland, Baltimore County
JAMES L. WESCOAT, JR., University of Illinois at Urbana-Champaign
GARRET P. WESTERHOFF, Malcolm Pirnie, Inc., Fair Lawn, New Jersey

---

[*] Terms expired June 30, 2006.

Staff

STEPHEN D. PARKER, Director
LAUREN E. ALEXANDER, Senior Program Officer
LAURA J. EHLERS, Senior Program Officer
JEFFREY W. JACOBS, Senior Program Officer
STEPHANIE E. JOHNSON, Senior Program Officer
WILLIAM S. LOGAN, Senior Program Officer
M. JEANNE AQUILINO, Financial and Administrative Associate
ELLEN A. DE GUZMAN, Senior Program Associate
ANITA A. HALL, Program Associate
DOROTHY K. WEIR, Research Associate
MICHAEL J. STOEVER, Program Assistant

Appendix D

Biographical Information for Committee on the Scientific Bases of Colorado River Basin Water Management

Ernest T. Smerdon (NAE), Chair, is an expert in water resources engineering and management, especially in the U.S. West. Dr. Smerdon is a retired former vice-provost and dean of the College of Engineering and Mines at the University of Arizona. Dr. Smerdon has served as an advisor to the U.S. federal government and several foreign governments on water resources development and agricultural issues. Dr. Smerdon has served on several National Research Council (NRC) committees, most recently serving as chairman of the committee that authored *Managing the Columbia River: Instream Flows, Water Withdrawals, and Salmon Survival* (2004). Dr. Smerdon is thus well versed in western water science and policy matters and the NRC study process. Dr. Smerdon received his B.S., M.S., and Ph.D. degrees, all in engineering, from the University of Missouri.

Julio L. Betancourt is a research scientist with the U.S. Geological Survey Desert Laboratory in Tucson, Arizona. Dr. Betancourt's research focuses on ecosystem and watershed responses to climate variability on different temporal and spatial scales. Dr. Betancourt employs various techniques and approaches to help reconstruct pre-instrumental hydrologic and climatic data. These include the use of rodent midden and tree-ring data in the Americas, and the design and testing of approaches that include historical documents and photographs, instrumental hydrologic and climatic data, long-term vegetation plots, tree rings, stable isotopes, ancient DNA, biometric measurements, alluvial stratigraphy, and ice core reconstruction. Dr. Betancourt received his B.A. degree in anthropology from the

University of Texas, Austin, and his M.S. and his Ph.D. degrees in geosciences from the University of Arizona.

Gordon W. "Jeff" Fassett, P.E., is the National Director for Water Resources at HDR Engineering, Inc. and is based in Cheyenne, Wyoming. Prior to joining HDR in 2006, he was president of Fassett Consulting LLC, where he specialized in water rights, water resources engineering, and water management and policy matters in the western states. Prior to opening his firm in 2000, Mr. Fassett served as the State Engineer of Wyoming from 1987 to 2000. In that post, he directed all policy, technical, and administrative issues of the cabinet-level state government agency responsible for the appropriation, beneficial use, and general supervision and regulation of all waters in the state. In addition, Mr. Fassett was Wyoming's representative for all of the shared interstate rivers and worked on a variety of river, reservoir, and environmental water management issues, including those within the Colorado River basin. Mr. Fassett received his B.S. degree in civil engineering from the University of Wyoming.

Luis A. Garcia is a professor of civil engineering at Colorado State University, where he specializes in the fields of irrigation and drainage, decision support systems for water resources decision making, and computer modeling and geographic information systems applications. Dr. Garcia has worked on these issues in many locales across the U.S. West, as well as in several foreign countries, including Egypt, India, and Italy. Dr. Garcia received his B.S. and M.S. degrees in civil engineering from Texas A&M University, and his Ph.D. degree in civil engineering from the University of Colorado.

Donald C. Jackson is a professor of history at Lafayette College in Easton, Pennsylvania. Dr. Jackson's areas of professional interest are in western U.S. water development and history, and the engineering and political aspects of large dam construction and operations. Dr. Jackson's books on these topics include *Big Dams of the New Deal Era: A Confluence of Engineering and Politics*, co-authored with David P. Billington (2006), *Building the Ultimate Dam: John S. Eastwood and the Control of Water in the West* (1995), and *Great American Bridges and Dams* (1988). Dr. Jackson has served as a fellow at the Dibner Institute for the History of Science and Technology at the Massachusetts Institute of Technology and as a fellow at The Huntington Library in San Marino, California. Dr. Jackson received

Appendix D

his B.S. degree in engineering from Swarthmore College, and his M.A. and Ph.D. degrees in American studies from the University of Pennsylvania.

Dennis P. Lettenmaier is a professor of civil engineering at the University of Washington. Dr. Lettenmaier's specialties are hydrologic modeling and prediction, hydroclimatology, and remote sensing. In addition to his service at the University of Washington where he has been a faculty member since 1976, he has served as a visiting scientist at the U.S. Geological Survey in Reston, Virginia (1985-1986) and as program manager of NASA's Land Surface Hydrology Program at NASA Headquarters (1997-1998). Dr. Lettenmaier was the founding Chief Editor of the American Meteorological Society's *Journal of Hydrometeorology*. He has authored or co-authored over 150 journal articles on topics ranging from the hydrologic impacts of climate change to Arctic hydrology. Dr. Lettenmaier received his B.S. degree in mechanical engineering from the University of Washington, his M.S. degree in civil, mechanical, and environmental engineering from George Washington University, and his Ph.D. degree in civil engineering from the University of Washington.

Eluid L. Martinez is the president of Water Resources Management Consultants LLC in Santa Fe, New Mexico. Prior to assuming his current position in 2001, Mr. Martinez served as the Commissioner of the U.S. Bureau of Reclamation from 1995 to 2001. Before his service with the Bureau, Mr. Martinez served as the New Mexico State Engineer from 1990 to 1994. Mr. Martinez thus has experience in working with several western U.S. river management agreements and compacts, including the Colorado River, Rio Grande River, and the La Plata River. Mr. Martinez received his B.S. degree in civil engineering from New Mexico State University.

Stephen C. McCaffrey is Distinguished Professor and Scholar at University of the Pacific's McGeorge School of Law and is an expert on international water resources law. He is a former chairman of the United Nations (UN) International Law Commission. During his time there he guided the work that formed the basis of the 1997 United Nations Convention on the Law of the Non-Navigational uses of International Watercourses, a treaty designed to ensure the equitable use of waters shared by more than one country. Dr. McCaffrey currently serves as legal consultant to the Nile River Basin Cooperative Framework project, a UN-sponsored effort to forge a basin-wide agreement among the 10 basin states on utilization of Nile water resources. He has argued before the World Court, advised the State Department, and represented foreign governments in river-use disputes. Dr. McCaffrey has published a number of books and more than 70 articles in law journals. He received his B.A. degree from the University of Colorado, his J.D. from the University of California, Berkeley, and his Dr. iur. degree from the University of Cologne, Germany.

Eugene M. Rasmusson (NAE) is research professor emeritus in the Department of Meteorology at the University of Maryland in College Park. Dr. Rasmusson's research interests are in the atmospheric general circulation and the global hydrologic cycle. Much of his research has centered on the relationship between sea-air interaction in the tropics and global precipitation variability, with particular emphasis on the El Niño phenomenon of the tropical Pacific. He also has interests in developing methods for improved prediction of climate variations and their impacts on water resources. Dr. Rasmusson received his B.S. degree in civil engineering from Kansas State University, his M.S. degree in engineering mechanics from St. Louis University, and his Ph.D. degree in meteorology from the Massachusetts Institute of Technology.

Kelly T. Redmond has served for 18 years as Regional Climatologist and 15 years as Deputy Director at the Western Region Climate Center, located at the Desert Research Institute in Reno, Nevada. He previously served 7 years as state climatologist for Oregon. Dr. Redmond's research and professional interests span every facet of climate and climate behavior, its physical causes and variability, methods and properties of measurement systems, how climate interacts with other human and natural processes, and how such information is acquired, used, communicated, and perceived. He received a B.S. degree in physics from the Massachusetts Institute of Technology, and M.S. and Ph.D. degrees in meteorology from the University of Wisconsin.

Philip M. Smith is a consultant with Science Policy and Management in Santa Fe, New Mexico. He has been involved in national and international science and technology policy and program development for more than four decades. Mr. Smith has periodically been engaged in water research programs and policy over these years. From 1981 to mid-1994, he was Executive Officer of the National Academy of Sciences and the National Research Council. For more than 20 years Mr. Smith was a government research management and science and technology policy official with the White House Office of Science and Technology Policy, Office of Management and Budget, and the National Science Foundation. He received B.S. and M.A. degrees from Ohio State University and was awarded an honorary D.Sc. degree by North Carolina State University.

Connie A. Woodhouse is an associate professor in the Department of Geography and Regional Development at the University of Arizona. Prior to joining the University of Arizona, she was a research scientist at the University of Colorado's Institute of Arctic and Alpine Research and a physical scientist within the National Climatic Data Center of the National Oceanographic and Atmospheric Administration. In her research, Dr. Woodhouse has focused on the generation and interpretation of high-resolution records of climate for the past 2,000 years. Her current research includes tree-ring reconstructions of drought for the Great Plains and Rocky Mountains, and investigations into the mechanisms of long-term drought and impacts on ecosystems. Dr. Woodhouse received her B.A. degree from Prescott College, her M.S. degree in geography from the University of Utah, and her Ph.D. degree in geosciences from the University of Arizona.

STAFF

Jeffrey W. Jacobs is a senior program officer at the NRC's WSTB. Dr. Jacobs' research interests include policy and organizational arrangements for water resources management and the use of scientific information in water resources decision making. He has studied these issues extensively both in the United States and in mainland Southeast Asia. Since joining the NRC in 1997, Dr. Jacobs has served as the study director of 17 NRC committees. He received his B.S. degree from Texas A&M University, his M.A. degree from the University of California, Riverside, and his Ph.D. degree from the University of Colorado.

Dorothy K. Weir is a research associate with the WSTB. She has worked on a number of studies including Everglades restoration progress, water quality improvement in southwestern Pennsylvania, and water system security research. Ms. Weir received a B.S. in biology from Rhodes College in Memphis, Tennsessee and an M.S. degree in environmental science and policy from Johns Hopkins University. She joined the NRC in 2003.